Cooperation

Among

Animals

Cooperation Among Animals

with human implications

A Revised and Amplified Edition
of THE SOCIAL LIFE OF ANIMALS

by **W. C. Allee** *Head Professor of Biology,*
The University of Florida
Professor of Zoology Emeritus,
The University of Chicago

Henry Schuman   New York

This book is gratefully dedicated to the past and present

members of our "Ecology Group"; without their enthusiastic

cooperation much of the underlying evidence could

not have been collected during my lifetime, and without their

critical attention the expression of these ideas

would have been more faulty

Contents

Plates

Figures

Foreword

I was honored in 1937 by an invitation to give the Norman Wait Harris lectures at Northwestern University; the more so since as one of their side-door neighbors I lived close enough for my personal foibles to be well known, thereby removing the chief source of any possible glamour. In this book that grew out of those lectures, as in the lecture series itself, I make no effort to pose as the remote purveyor of a mysterious erudition; I could not in any case regard myself as more than the exponent of the glorified common sense that I more and more firmly believe all science should be.

Even more than most, this book is the outgrowth of years of cooperative effort. Some of the basic facts were collected with the help of funds from the Rockefeller Foundation given to aid biological research at the University of Chicago. Other researches were supported directly by that university and by a grant from the National Research Council for the study of the effect of hormones on behavior.

In addition to the personal aid received from my scientific associates, many of whom will be named in the text, the kindly criticism of Professor Alfred E. Emerson has been particularly helpful in developing the work and in shaping the content and implications of these lectures. My thanks are given also to Professor Sewall Wright for his criticism of Chapter VI, to Mr. Kenji Toda for preparing the illustrations, and to the late Marjorie Hill Allee, whose command of the written word was a most helpful resource.

W. C. ALLEE

The University of Chicago
The University of Florida
  Gainesville, Florida

Cooperation Among Animals

The Approach.

Chapter I

The rate of obsolescence of material things is high. This is particularly true of consumers' goods; and even capital goods usually become out of date in a long generation. In 1937 an admirer of Will Rogers dedicated a lasting monument to the humorist. Although built for time and erected in our semi-arid West, where decay is slow, the tower is expected to last only a thousand years. Invested capital evaporates even with watchful care; there are few private collections of material wealth that remain intact a third of a thousand years.

The most permanent contributions of our age appear to be the scientific discoveries we have made, the artistic beauties we have created, and the ideas we have evolved. To the extent that these advances are entombed in libraries and museums they share the impermanence of more material things. A nearer approach to immortality is permitted those bits of science and art that escape from the bindings of books and pass into the active life and traditions of people. The more widespread and firmly fixed these become in the minds of living men, the greater is their chance of longevity.

The most practical achievement of our extremely practical period is the habit of searching for new truths and for correct interpretations of those long known. The unique contribution of the present era is not that made by men of business and affairs, spectacular as it may be. Rather this age is and will be known as the time of the development and application of scientific methods. This contribution is being made by extremely impractical research workers who are supported by a tiny splinter from the great block of capital gains. Money spent effectively to this end, whether in the aid of research or other creative scholarship or in teaching the results gained by research, makes the most lasting and important of all modern investments. The most nearly permanent monument any man can erect is to influence directly or indirectly the growth of improved ideas and traditions among the men in the street, in the factory, or on the farm.

It is in this spirit that I have undertaken to interpret one of the significant biological developments of recent years. It is my hope that through the work described in these pages all social action may be given a somewhat broader and more intelligent foundation.

We may gain the impression from some modern over-simplifications that science deals with mere facts, that philosophy attends to principles and eternal truths, and that religion is concerned with values. In the following pages it will be necessary to shake aside such artificial limits and to present principles along with the evidence that supports them; to test these against experience and to attempt frequently to weigh the general biological values involved. This last process will be easier if we assay survival values only. Admittedly, in dealing even with survival values we must be relatively rough and ready in our methods, and perhaps the conclusions will carry a strong odor of the laboratory in which they had their origin.

Basically, the approach will be that of the experimental biologist rather than that of the theorist—which might be more polished—or of the philosopher—which would certainly be more abstract and would probably use a great many more words for the same number of ideas. Despite much belief to the contrary, any biological fact that concerns us can be accurately described and the conclusions from its study clearly expressed in relatively simple and direct language.

, In research reports and scholarly discussions there is need for the conciseness and precision made possible by technical language. Science is ill-served, however, by any tendency to develop a cult of obscurity. Scientists must be free to attack the unknown as effectively as they can, and in return for intellectual freedom they have an obligation to interpret research results in terms that can be understood by intelligent and interested people.

There is current in the United States at least one major attempt to organize all knowledge about metaphysics, and so secure a longed-for unity. In order to obtain a simplified system, the group of men occupied with this enterprise turn back to the days before the present scientific era to find a statement of eternal principles that will serve as a unifying nucleus for human experience and thought. Such efforts, while furnishing an excellent corrective for overconfident scientists, seem mischievously naïve as a serious, present-day movement. We do need relief from our absorbed attention to conflicting scientific detail, but progress must come from newer syntheses that take account of the world and man as science sees them rather than by accepting almost as a whole some ancient system of historical significance. These systems are out of date primarily because they were evolved before one of the greatest advances in knowledge that man has yet been able to make, that of modern science.

Modern philosophical educational systems, if they are to survive, must have as their central core the well-tested evidence compiled by objective scientific methods. Such knowledge must have stood the test of being checked and rechecked by men constitutionally agnostic in their mental attitudes, who can say, "I do not know. What is the evidence?"

An anecdote that is becoming classic among scientists will illustrate the point. Professor Robert Wood, physicist of the Johns Hopkins University, was asked to respond to the toast "Physics and Metaphysics" at a dinner of some philosophical society. His response was somewhat as follows: The physicist gets an idea which seems to him to be good. The more he mulls over it the better the idea appears. He goes to the library and reads on the subject and the more he reads the more truth he can see in his idea. Finally he devises an experimental test and goes to his laboratory to apply it. As a result of long and careful experimental checking he discards the idea as worthless. "Unfortunately," Professor Wood is said to have concluded, "the metaphysician has no laboratory."

History.

Chapter 11

Like other human disciplines, science has its orthodox and its heterodox views. The idea that unconscious automatic cooperation exists has had a long history, and yet it is just now beginning to escape from the heterodox category.

My own interest in this subject dates not from a preconceived idea but from a clearly remembered bump against some stubborn experiments. Almost forty years ago as a graduate student in zoology I was engaged in studying the behavior of some common small fresh-water animals called isopods, tiny relatives of the crayfish. All fall and winter I had been collecting them from quiet mud-bottomed ponds, chopping the ice if necessary, and from beneath stones and under leaves in clear small streams.

I kept them in the laboratory under conditions resembling those in which they lived in nature. Then day after day I put lots of five or ten isopods into shallow water in a round pan that had a sanded wax bottom so prepared that the isopods could crawl about readily. When a current was stirred in the water the isopods from the streams usually headed against it; but those from ponds were more likely either to head down current or to be indifferent in their reaction to the current. The behavior of the two types was sufficiently different so that at first I thought that stream and pond isopods represented different species, but the specialist at the National Museum assured me that all belonged to the species appropriately called *Asellus communis*, the commonest isopod of our inland waters.

Rather cockily I reported after a time to my instructor that I had gained control of the reaction of these animals to a water current. By the judicious use of oxygen in the water, I could send the indifferent pond isopods hauling themselves upstream, or I could reduce the stream isopods to going with the current. I had not reckoned with another factor that presently caught up with me.

After a winter in the laboratory it seemed wise as well as pleasant to take my pan out to a comfortable streamside one sunny April day and there check the behavior of freshly collected isopods in water dipped from the brook in which they had been living. To my surprise, the stream isopods, whose fellows all winter had gone against the current, now went steadily downstream or cut across it at any angle to reach another near-by isopod. When I used five or ten individuals at a time, as I had done in the laboratory, they piled together in small close clusters that rolled over and over in the gentle current. Only by testing them singly could I get away from this group behavior and obtain a response to the current; and even this reaction was disconcertingly erratic.

It took another year of hard work to get this con-

tradictory behavior even approximately untangled;[1]* to
find under what conditions the attraction of the group
is automatically more impelling than keeping footing in
the stream; and that was only the beginning of the road
that I have followed from that April day to this time,
continuing to be increasingly absorbed in the problems
of group behavior and other mass reactions, not only of
isopods, but of all kinds of animals, man included.

As the years have gone on, aided by student and other
collaborators and by the work of independent investigators, I have tried to explore experimentally the implications of group actions of animals. Necessarily, I have had
to turn to the world's literature to find what others
have done and are doing along this line.

A Greek philosopher named Empedocles seems to have
had the first recorded glimmerings of an idea of the universal and fundamental nature of the cooperation that
underlies group action, as well as a conception of the opposing principle of the struggle for existence. He lived in
the fifth century B. C., and he was not only a thinker
but so much a man of affairs that he was offered a king's
crown, which he refused.[142]

Empedocles owes his present-day fame to two long
poems in which he outlined the idea that there are
natural elements—fire, earth, air and water—which are
acted upon by the combining power of love and the dis-
rupting power of hate. Under the guidance of the building force of love the separate elements came together
and formed the world. Separate parts of plants and various unassorted pieces of animals arose from the earth.
These, Empedocles taught, were often combined and at

* Detailed citations to more complete statements will be
found in the bibliography.

first the results were monstrous shapes, which in time became altered until, still guided by combining love, they clicked, to make the more perfect animals we now know. It has taken us almost two and a half millenia to transmute this poetic conception into the less picturesque but more exact and workable expression acceptable to modern science.

After the fertile Greek era there intervened in this field as elsewhere the long sterile period when Greek philosophy, if known, was dogmatically accepted, and shared with other authoritarian systems the responsibility of explaining the world of reality as well as the universe of fancy.

It was not until my own experiments and thinking and reading had begun to form in my mind a fairly definite pattern that, by the aid of Havelock Ellis' *The Dance of Life*,[48] I stumbled upon the ideas of the third Earl of Shaftesbury, who lived before and after 1700. He seems to have been the first intellectual in the modern period to recognize fairly clearly that nature presents a racial impulse that has regard for others, as well as a drive for individual self-preservation; that, in fact, there are racial drives that go beyond personal advantage and can only be explained by their advantage to the group.

An unfriendly contemporary wrote pretty much these words: "Shaftesbury seems to require and expect goodness in his species as we do a sweet taste in grapes and China oranges, of which, if any are sour, we boldly proclaim that they are not come to their accustomed perfection." Havelock Ellis, in reviewing this development, says that "therewith 'goodness' was seen practically for the first time in the modern period to be as 'natural' as the sweetness of ripe fruit." It is only fair to record that in the religious world for at least fifty years previous there had been growing a similar conviction among certain heretics.

Adam Smith, in his *Theory of Moral Sentiments*, wrote in 1759 about the same qualities under the heading of "sympathy" or "fellow feeling." His more famous *Inquiry into the Wealth of Nations* (1776) is based on the opposed forces of self-interest, and he did not publicly reconcile the two. Later, Comte,[40] who died in 1857, emphasized "altruism," as Feuerbach,[51] of the same general period, did "love." These relationships were viewed sympathetically by Lange (1875) in his remarkable *History of Materialism*.[84]

Herbert Spencer argued now for egoism and again for altruism. In 1901 he balanced the two in the following quotation from his *Principles of Ethics*: "If we define altruism as being all action, which in the normal course of things benefits others instead of benefiting self, then, from the dawn of life, altruism has been no less essential than egoism. Though primarily it is dependent on egoism, yet secondarily egoism is dependent on it."

Charles Darwin in the *Origin of Species* recognized that evolution within the worker caste of ants can be explained by the fact that selection can act on the family as well as on the individual. Such ideas are in keeping with Darwin's use of the phrase "struggle for existence," of which he said in the *Origin*: "I use this term in a large and metaphorical sense including dependence of one being on another and including (which is more important) not only the life of the individual, but success in leaving progeny."

In the *Descent of Man* (1874), Darwin gave naturalistic examples of mutual aid; and we can readily see that his whole thesis that man is descended from other animals requires that man's altruistic drives have precursors among his animal ancestors. Darwin clearly derives the moral sensibility of man from his social drives played upon by reflective intelligence; he states: "The term, general good, may be defined as the rearing of the greatest number of individuals, in full vigour and health, with all their faculties perfect, under the conditions to which they are subjected. As the social instincts both of man and the lower animals have no doubt developed by nearly the same steps, it would be advisable, if found practicable, to use the same definition in both cases, and to take as the standard of morality, the general good or welfare of the community. . . ."

"It may be regretted," Keith, the English anthropologist, wrote in 1949,[81] "that Darwin did not lay greater emphasis on the part played by co-operation in his scheme of evolution." There is no doubt that the general tone of Darwinism has taken color from the extreme individualism of Darwin's time.

In 1930, after having written the text of a careful account of experimental evidence concerning the existence and nonexistence of cooperation at subsocial levels,[3] I set down in the draft of a proposed preface the idea that the existence of such a principle was now for the first time an established fact, for which the details to follow gave full proof. I still think that the proof is good. However, the preface as published does not contain any such claim of priority, for at that point in the writing I went back and reread *Des sociétés animales*, by the French scientist Espinas,[49] which appeared in 1878 and which is the pioneering essay in this field so far as modern work is concerned. There I found Espinas affirming that no living being is solitary, but that from the lowest to the highest each is normally immersed in some sort of social life; he added that he was ready to offer conclusive proof.

I turned the pages of Espinas' detailed history of the evolution of ideas about the origin and development of society and came to his massed evidence that communal life is not "a restricted accidental condition found only among such privileged species as bees, ants, beavers, and men, but is in fact universal."

The evidence was based largely on observations of the existence in nature of animal groupings, which are found widely distributed in the different levels of the animal kingdom. It was clear to me that the facts which Espinas had found so impressive had not convinced others; and although they were suggestive, they did not seem compelling to me in the light of other indications to the contrary. Perhaps, I cautioned myself, even the experimental evidence that I had accumulated in 1930 was not really crucial, and it would be well to avoid making too strong a claim in the matter. The same caution must continue even in the face of still stronger evidence on hand today.

Coming in 1878, the conclusions of Espinas are the more important because the scientific world was then, as men in the street are today, under the spell of the idea that there is an intense and frequently very personal struggle for existence which is so important and far-reaching as to leave no room for so-called "softer" philosophies. The idea of the existence of natural cooperation was apparently in the air despite the preoccupation with the egoistic phase of Darwinism. Kessler[6] addressed a Russian congress of naturalists on this subject in 1880, and from this address sprang the remarkable if uncritical book on mutual aid by the Russian anarchist Prince Kropotkin (1902).[83]

By combing the accumulated natural-history records, Kropotkin was able to collect observation after observation indicating that animals in nature do aid each other to live, as well as, on occasion, kill each other. Kropotkin's work served the admirable purpose of keeping this idea alive and popularizing it. It has had also the less fortunate result of bringing Kropotkin's fundamental doctrine into disrepute among students who are critically sensitive to the value of evidence and who find that Kropotkin's sources were not always reliable.

But I have gotten ahead of my story. In 1889 (and again in 1911)[60] Geddes and Thompson concluded that general progress and great advances among both plants and animals must be regarded not simply as individual achievements "but very largely in terms of sex and parenthood, of family and association; and hence of gregarious flocks and herds, of cooperative packs, of evolving tribes, and thus ultimately of civilized societies. . . ." Still the intellectual climate was unfavorable and the idea of cooperation did not catch general attention even among biologists, despite the emphasis placed upon it by several other competent writers. As a confirmed experimentalist, I am inclined to think that the lack of experimental evidence was fatal. However, it may turn out that the idea will not catch on now, even though much experimental evidence has been collected in the last thirty years.

William Patten, an American biologist who taught for many years at Dartmouth College, made the next general statement of the fundamental nature of cooperation when in 1920 he gave it a central place in his analysis of the grand strategy of evolution.[101] It is of personal interest to me that, at the scientific meetings in 1919 at which I presented my first experimental results on this subject, Professor Patten gave a vice-presidential address in which he outlined, mainly from philosophical considerations, his conclusions concerning the importance of biological cooperation. He was rightly impressed by the fact that cells originally were separate, as protozoans are today. Some, however, evolved the habit of remaining attached to one another after division. This made a beginning from which the many-celled higher animals could develop. With each increase in the ability of cells to cooperate there came power to increase the complexity of organization of the cell masses. The highly evolved bodies of men are thus an expression of increasing intercellular cooperation that finally reaches a point at which, for many purposes, the individual person becomes the unit rather than the individual cells of which he is composed.

About the same time Deegener,[45] a German, published an extensive treatise on the social life of animals, along the same lines as the book written by Espinas forty years before. Deegener's distinctive contribution was a classification of the different social levels, from the simplest sorts of artificial collections of animals to parasitism and truly social life. His rating of these different aspects of subsocial and social life in one long outline has the great merit of showing that no hard and fast lines can be drawn between social and subsocial organisms, but that social communities are the natural outgrowth of subsocial groupings. Unfortunately, with Teutonic vigor and vocabulary, he designated the different categories in words as unwieldy as they were exact. Bogged down by the weight of such terms as *sympatrogynopaedium*, *synaporium* and *heterosymphagophaedium*, Deegener's real contribution tends to be lost even to biological scholars.

A survey such as I am attempting here should not try to be exhaustive; I shall dismiss with a word the slight advance made by Alverdes[20] and shall make no mention whatever of the work of many others. There is, however, a phase of the literature which has given me so much pleasure as well as useful information that I shall not pass it over: this phase deals with the social insects. Espinas, Kropotkin, Deegener, and Alverdes, of those already mentioned, and a host of others have written in detail and in general about these fascinating insects, but none has written more accurately or with greater insight and literary as well as scientific skill than the American entomologist William Morton Wheeler. His book *Social Life Among the Insects,* which appeared in 1923, is a noteworthy general summary.[132] In this he shows that among insects alone, including such well-known forms as termites, bees, wasps, ants, and the less generally known social beetles, the social habit has arisen some twenty-four distinct times in about one fifth of the known major divisions of insects. It would seem that there is a general reservoir of presocial traits from which, given the proper opportunity, society readily emerges. Wheeler, no less than Espinas, from whom he quotes, emphasizes that even so-called solitary species of animals are of necessity more or less coöperative members of associations of animals and that animals not only compete among themselves but also cooperate with each other to secure mates and insure greater safety.

However, the fact that Wheeler drew his illustrative material primarily from, and based his conclusions mainly on, his knowledge of social life among insects did not make for the full acceptance of his ideas. The existence of cooperation among nest mates in ants and bees does not prove that there are beginnings of coöperative processes among amoebae and other greatly generalized animals.

Man and the few species of highly social insects are a small part of the animal kingdom; in order to discover and distinguish the principles of general sociology it is necessary to look farther, to focus attention on the social and antisocial relationships of many animals usually regarded as lacking social life.

With and without this end in view, simultaneous but independent outbreaks of experimentation on group effects among the lower animals have been occurring in the last several decades. For a time just preceding and following 1920 we who in Australia,[117] in France,[30] and in the United States[2] were engaged in these studies, continued unaware of each other's work. Relatively soon, however, since biological literature is today widely and, for the greater part, promptly circulated, all such work, even that in Russia,[59] became generally known. It is these general experiments on population growth, on mass physiology, and on animal aggregations that are now the important aspect of the field of proto-cooperation—which may be defined as the beginnings of cooperation—or of full animal cooperation.

Natural History.
Chapter III

In the preceding chapter I have briefly traced the history of the idea of innate cooperation. One reason for the slowness with which that idea had been accepted is the obvious fact that cooperation is not always plain to the eye, whereas competition in its most noncooperative form, in which no positive social values are apparent, can readily be observed. With certain exceptions that will soon be mentioned, it has seemed that, social species aside, crowding, the simplest start toward social life, which is easily apparent and a condition of nearly all society, is harmful to the individual and to the race alike. Since 1854[70] it has been known from experimental evidence that crowded animals may not grow at all and when they do grow do so less rapidly than their uncrowded relatives. Under many conditions crowded animals not only do not grow but they die more readily and frequently reproduce less rapidly than when living in uncrowded populations.

All the older works in natural history taught fairly clearly that crowded groups, to have real survival values, must be sufficiently well organized to contribute to group safety by warning of danger or by defense in case of attack.[3] If, in addition, these groups are organized on a basis of division of labor, such as occurs in the highly social colonies or ants and termites, with specialized reproductives, workers, and soldiers, or according to the patterns found in human society, then the survival values of groups are readily seen.

Yet, for some reason, under natural conditions and with very many sorts of animals, crowding in all degrees does occur and apparently always has occurred. Conceded that animals do not always act for their own best interests, still they must do so to a certain degree or be exterminated in the long run. The advantages of the long-established habit of a species may not be obviously apparent, but it is not safe to say offhand that advantages do not exist.

There are, for example, dense crowds of certain animals, such as ladybird beetles (Plate I, top), that with the approach of winter collect in restricted and favorable places where they hibernate together. Apparently, in the face of winter cold, there is some safety in numbers even among cold-blooded animals that collect in hordes without any organization.

A second plain exception to the general testimony that crowding of nonsocial species is harmful is to be seen in the aggregations that form during the breeding season. Like the hibernating groups, these are very widely distributed through the animal kingdom. Breeding aggregations of worms, crustaceans, fishes, frogs, snakes, birds, and mammals—or the midge insects such as those shown in Plate I, bottom, for example—have long attracted attention. Such aggregations have been numerous enough and conspicuous enough to stimulate repeated descriptions by naturalists.

A third exception is found during times of migration, when animals frequently crowd together in great hordes and execute mass migratory movements, like those of many birds.

Breeding, hibernation, and migration aside, the older information indicated that, up until the point that social life is developed, crowding is harmful, but there are many other instances of crowding that do not fall within any of these classifications. It will be worth while to consider the extent and the natural history of some of these dense animal aggregations. Here, as elsewhere, there will be no attempt to catalogue all known instances or to select merely the very best cases known. I shall try to use examples that are not too shopworn by repeated description.

Almost every observant person has seen the soft green "bloom" which covers many stagnant ponds. Under the microscope this "bloom" is often seen to be composed of myriads of the tiny plant-animal *Euglena*. These organisms are commonly one tenth of a millimeter long, which means that in a characteristic layer of "bloom" there would be at least sixty to one hundred thousand animals per square inch; and acres of water are sometimes covered.

Lobster-krills are small crustaceans that occur commonly in shoals about the Falkland Islands, Patagonia, New Zealand, and other southern waters.[91] A larval stage of this animal, less than an inch long, occurs often on the surface of the water in such numbers that the sea is red for acres; and whales in those waters simply open their mouths and swim through slowly, feeding with no more effort than it takes to strain out the animals. These shrimp-like animals may be piled up on the shore by tide

PLATE I. (above). Ladybird beetles collect in dense aggregations in the autumn and hibernate. (below). During their breeding season, male midges gather in swarms and await the coming of the females. (Photographs by Welty.)

and wind in stench-producing layers. Dampier wrote of them in 1700: "We saw great sholes of small lobsters, which colored the sea red in spots for a mile in compass"; and they have been known to extend along the Patagonian coast for as much as three hundred miles.

At Woods Hole, on Cape Cod, I have at certain seasons dipped up a bucket of sea water from the harbor and found more space occupied by clear, jelly-like ctenophores, each the size of a walnut, than was taken by water. Sometimes I have dipped up a fingerbowl of sea water and found it so filled with small pin-point-like copepods that again there seemed to be more of them than of the water itself. These tiny relatives of the lobsterkrills also are the food of whales, and they, too, may discolor the ocean for miles.

Around bodies of fresh water, May flies or midges may emerge in clouds. At Put in Bay, near the lights flooding the monument that commemorates Perry's victory, I have picked up living May flies by the double handfuls from the millions that fly toward the lights; and near by our lake boat steamed through windrows of cast skins of the emerging May fly nymphs. Nearer Chicago I have taken water isopods, the half-inch crustacean mentioned earlier, by the bucketfuls from pools where they had collected in numbers only to be compared with those in twenty swarms of bees.

We have already spoken of the migratory hordes. Locusts in migration[128] swarm out of the sky in the Sahara borderlands, in southern Russia, in South Africa, and on the Malay Peninsula in terrorizing numbers (Figure 1). They once did so on the Great Plains of the United States, leaving a lively memory of destruction that is still roused by the smaller migrations that may occur there any summer in spite of active control measures. I myself have seen the so-called Mormon cricket advancing from the

relatively barren mountain pastures of Utah into the green
fields in numbers which were not halted by the hawks,
turkeys, and snakes attendant on the swarm and feeding
greedily; or by the active assaults of men and children
warned out to protect the cultivated lands. Migrating
army worms and chinch bugs present equally impressive
aggregations.

The emergence of Mexican free-tailed bats from the
Carlsbad cave of an August evening has been described
as a black cloud pouring out in such density as to be visible two miles away.[23] In the winter such bats are estimated to hibernate in these caves by the millions; and

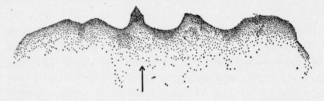

FIGURE 1. *A band of grasshopper nymphs on the march.
(From Uvarov, by permission of the Imperial Bureau of
Entomology.)*

they may be found through the day in sleeping masses
a yard across, hanging from the roof like swarms of bees.

Even larger mammals may collect into great, closely
packed herds. The migrating caribou on the tundra
move south in autumnal hordes that flow past a given
point for hours or even for days. And of the antelope
on the plains of Mongolia,[21] Roy Chapman Andrews
says that he has seen thousands upon thousands of bucks,
does, and fawns pour over the rim and spread out on the
plain. Sometimes a thousand, more or less, would dash
away from the herd, only to stop abruptly and feed.
The mass of antelope were in constant motion even when

they were undisturbed. They scattered before his auto-
mobile only to re-form within a few hours. In that region
only the grassland antelope gathers in such immense
herds; the long-tailed desert species never does so, probably because there is not enough food to support them
in their more arid dwelling place.

These are merely a few of the more dramatic instances
of the collection of great masses of animals in a small
space. They are more spectacular but probably less important than are the innumerable smaller aggregations
of animals that are frequently encountered. The small
dense crowds of whirligig beetles are a case in point.
These occur in widespread abundance on the surface of
our inland waters.

The more common condition of less intense crowding
does not mean that animals are usually solitary. Rather
the growing weight of evidence indicates that animals are rarely solitary; that they are almost necessarily members of loosely integrated racial and interracial
communities, in part woven together by environmental
factors and in part by mutual attraction between the
individual members of the different communities, no one
of which can be affected without changing some or
even all the rest, at least to some slight extent. Con-
tagious distribution is the rule in nature.

Let us take an example. Before the coming of the
white man, and even a century ago or less, much of the
Great Plains was occupied by what ecologists call a
grassland-bison community. Grasses readily grew in
the rich soil, despite the usual summer dry spells and the
more severe cyclic droughts that occurred even then.
By keeping the grasses fairly closely cropped the bison
herds prevented the invasion of herbs and shrubs that
might have withstood the severities of the climate but
could not make headway against continual grazing (Plate

II). In this function the bison were joined by a myriad of grasshoppers, crickets, meadow mice, and prairie dogs. All these were key-industry animals. That is to say, in one way or another they converted the grass into meat of different sorts, on which the plains Indians, buffalo wolves, hawks, owls, and coyotes fed. If the grass failed, then, as today in the absence of bison, many of the key-industry herb-eaters and those that in turn fed on them must either starve, migrate into another community where they would be disturbing factors, or change their source of food and thereby disturb the balance in their own community.

It must be pointed out here that the plants of this community cannot be set off as separate from the animals. They divide the available space between them; they constantly interact upon each other and upon their physical environment; except for purposes of formal study or in limited fields, the biologist must consider both as members of a given association.

In such a community the effects of the dominant bison were felt in times of stress by the humblest and least conspicuous grasshopper. In the spring of the year hundreds of square miles normally supported populations of six to ten million insects and other invertebrate animals for every acre of land. As with warmer weather the predatory animals returned to the grasslands, these insects were eaten off until perhaps a tenth of their number could be found later in the season; with the autumn lushness they increased again, only to fall back to some half million or so per acre during the winter cold.

Similar communities exist among aquatic forms. In fact, one of the first demonstrations of such a community was made with the animals living in and on an oyster bank.[92] A beautiful and penetrating description of the interrelations that may be found in a small lake was

PLATE II. *A grassland-bison community. (Photograph from the National Park Board of Canada.)*

published not long after by the late Professer S. A.
Forbes,[54] of the Illinois Biological Survey, in which he
pointed out that minnows competed with bladderwort
plants for key-industry organisms; he showed too that
when a black bass is hooked and taken from the water the
triumphant fisherman is breaking, unsensed by him,
myriads of meshes which have bound the fish to all
the different forms of lake life.

The existence of these communities is now generally
recognized, and in order that they may exist it seems
that there must be a far-reaching, even if vague and
wholly unconscious, proto-cooperation among all the living creatures of the community. It is to such relationships that Wheeler referred when he said, "Even the
so-called solitary species are necessarily more or less cooperative members of groups or associations of animals
of different species." [132]

Within these communities aggregations of animals occur for a variety of reasons. Their nature can best be
shown by a series of illustrations.

One variety of aggregation is that of colonial forms,
in which many different so-called individuals remain
permanently attached together throughout their lives.
In the simplest cases all the individuals are alike. Each
possesses a mouth and food-catching tentacles, and each
feeds primarily for itself, although food caught by one
individual may be shared unconsciously with others near
by. In more complex forms some individuals have the
mouths suppressed, and receive all their food from those
that do catch food. They have become specialized as bearers of batteries of stinging cells; they strike actively
when the colony is touched, and their stinging cells ex-
plode so effectively as to give protection to the colony.
Other individuals in the same colony bear medusa-like
heads that break away and swim off, producing eggs and

sperm and distributing them as they drift. Here is certainly a division of labor, although these colonial animals are never rated as social.

Various modifications of such colonial animals are found particularly among colonial protozoa, sponges, and coelenterates; they also occur higher in the animal kingdom, even among the lower chordates, members of the same great phylum to which man himself belongs. It is interesting that animals whose structure forces them to the sort of compulsory mutual aid that automatically follows such structural continuity have never progressed far either in social achievement or in the evolutionary scale. When higher animals, such as the lower chordates, show this development, they are usually regarded as degenerate members of their general stock. These colonial animals are seldom dominant elements in the major communities of which they are a part. One comes to the conclusion that the more nearly voluntary such cooperation is, the greater its advantage in social life. It might on the other hand be pointed out that when an animal has achieved social organization and division of labor low in the evolutionary scale, the resulting colonies are so well adapted to their environment that there is not sufficient pressure to cause evolutionary changes.

A second type of aggregation occurs when animals are forced together willy-nilly by the action of wind, tidal currents, or waves over which they have no control and the effects of which they cannot resist. Many of the animal masses that lend color to wide patches of the ocean surface are brought together by temporary or permanent currents. Often animals so distributed are thrown down more or less by chance on types of bottom on which they can develop, and there, if favorable niches are somewhat rare, dense aggregations may result, like

New England coral on a suitably hard bottom, or the animals found on a wharf piling.

These accidental animal groupings may persist only as long as the physical forces which brought them together continue to act. Usually, however, they last somewhat longer, as a result of a slight inclination toward social inertia, which tends to keep animals concentrated in whatever place they happen to be found. If the groupings are to have much permanence, this quality of social inertia, the tendency of animals to continue repeating the same action in the same place, must be reinforced by another quality: the social force of toleration for the presence of others in a limited space. The densely packed communities of animals on a wharf piling can persist only if toleration for crowding is well developed.

Other dense collections may be brought about by forced movements of animals in response to some orienting influence in their environment. Some examples are the moths, June beetles, or May flies that collect about lights. Such aggregations are a result of the inherited, internal organization of the animals; and the irresistible attraction of the May fly to the light is joined with active toleration for the close proximity of others.

Other close aggregations occur as a result of the less spectacular trial and error reactions, in which the animals wander here and there, more or less vaguely stimulated by internal physiological states or external conditions, and so come to collect in favorable locations. Collections of animals about limited sources of food offer a good illustration. These, too, may show only the social qualities of inertia and toleration.

A decided advance is attained when animals react positively to each other and so actively collect together, not primarily because the location is favorable or through environmental compulsion but as the result of the beginnings of a social appetite. In early stages of such reactions, the movement together may come primarily because the collection of isopods or earthworms or starfishes is a substitute for missing elements in the environment.

Take, for example, the snake or brittle starfishes of the New England coast. Although these are now rare along Cape Cod, before the wasting disease swept away the eel grass they were abundant in favorable localities, but were rarely found close together. I have spent hours peering down through a glass-bottomed bucket here and there and round about in one of these localities, and have not seen more than one at a time. And I have spent more hours wielding a sturdy garden rake in swathe after swathe through the short eel grass, very rarely pulling in more than one starfish at a haul.

Yet when a few brittle starfishes are placed in a clean bucket of sea water they clump together like magic (Plate III). In bare laboratory aquaria they remain so clumped for weeks; in fact, the aggregations become more compact as time goes on as the animals bend back their extending arms and tuck them into the mass. If, however, the aquaria are dressed up by the introduction of eel grass so that conditions approach those found in nature, the aggregations disperse and the starfishes climb actively about over the blades of the eel grass, feeding on organisms and debris found on their surfaces.

The idea that in clean laboratory dishes these starfishes are substituting each other for the missing eel grass was obvious and easy to test. A kind of artificial eel grass was made of glass rods twisted in various shapes so that they offered a supporting framework for climbing in much the same way as does the true eel grass. So long as the rods remained, the starfishes clambered about over the mesh-

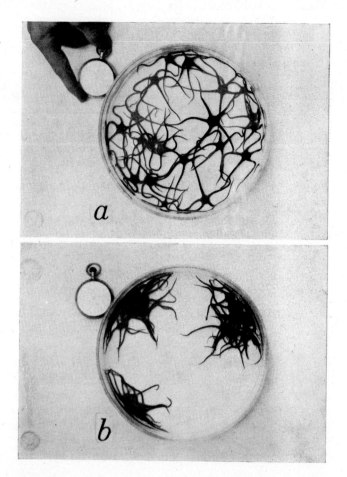

PLATE III. a. *Brittle starfish aggregate readily when put into a bare vessel of sea water.* b. *shows conditions ten minutes after a was taken. (Photographs by Welty.)*

work or hung motionless, usually isolated. When the rods were removed they again clustered together.

It is a far cry from such aggregations to the groupings of foreigners in a strange city that result in Little Italy, or the Mexican settlement, or a German quarter; and yet basically some of the factors involved are similar. Perhaps there is a closer connection between such aggregations in the wide expanse of a clean aquarium and the schooling tendency found among many fishes of the open sea; perhaps the same phenomenon accounts for the flocking tendency of many birds, as well as mammals on the equally monotonous grassy expanses of temperate plains.

A somewhat different expression of a positive social reaction is shown when animals that are usually more or less isolated come together and pass the night grouped as though they were engaged in a slumber party. This type of behavior has been repeatedly described for different insects, even for the wasps that remain separate to such an extent that they are called solitary wasps. In some forms of solitary wasps both males and females are found in the sleeping group. With solitary bees, such as we have near Chicago, the overnight aggregations are composed of males only. A study which was made of the sleeping habits of a Florida butterfly species indicates that these Heliconii[77] come together night after night in the same location, in part at least as a result of place-memory. The assemblages lack sexual significance. There is some protection in the fact that if one is disturbed the whole group may be warned. The presence of many butterflies would reinforce any species odor that might attract others of the same species or repel possible predators. Such groups of course have their dangers. The aggregated animals, apparently sleeping, provide a collected feast for the lucky predator that finds them.

The crowded roosts to which certain birds return not only for one season but sometimes for years are widely known. Here again we are concerned with a positive social appetite that grows stronger with the approach of darkness; the details as to why and how it operates are not known.

Animals that come together in intermittent groupings like these overnight aggregations are showing a social appetite that is none the less real because it is effective only at spaced intervals. In this it resembles other appetites, such as those for food, water, and sex relations. From such occasional or cyclic expressions of a social appetite it is a relatively short step to whole modes of life that are dominated by a drive for social relationships. In the insects alone this step has been taken some twenty-four distinct times and in widely separated divisions of that immense group.

Normally the development of highly social life comes by way of an extension of sexual and family relations over greater portions of the life span. Here again all degrees of increased length of association can be shown, from the sexual forms that meet but once and for a brief moment to the termite kings and queens that live together for years. Also all stages exist in the evolution of the association of parents with offspring, from insects such as the female walking stick, which deposits eggs as she moves about and pays no more attention to them, to the ants and bees whose worker offspring spend their entire lives in the parental colony or some colony budding off from it.

Athough the extension of family relations is very obviously one potent method by which social life is developed to a high level, there are other methods that also deserve consideration. Schools of fish arise, for example, under conditions in which there is no association with either parent after the eggs are laid. At times the eggs may be so scattered in the laying that the schools form from unrelated individuals. Here the schooling tendency seems to underlie rather than to grow out of family life. The mixed flocks[26] of tropical birds that are composed of many species obviously did not grow directly from family gatherings, and the groups of stags of Scottish deer, probably the original stag parties,[43] appear to give evidence of a grouping tendency independent of intersexual or family relations. This subject will be discussed in more detail in a later chapter.

The conclusion seems inescapable that the more closely knit societies arose from some sort of simple aggregation, frequently, but not necessarily, solely of the sexual-familial pattern. Such an evolution could come about most readily with the existence of an underlying pervasive element of unconscious proto-cooperation, or automatic tendency toward mutual aid among animals.

In the simpler aggregations, evidence for the presence of such cooperation comes from the demonstration of survival values for the group. These are more impressive the more constant they are found to be. If they exist throughout the year they are much more important as social forerunners than if present only during the mating season or at times of hibernation.

Beginnings
of Cooperation.
Chapter I V

With this chapter I begin the presentation of the evidence for the assertion that a general principle of automatic cooperation is one of the fundamental biological principles. The simplest expression of this principle is often found in the beneficial effects of numbers of animals present in a population. Laboratory work of the last few decades still shows that overcrowding is harmful, but it has also uncovered a no less real, though somewhat slighter, set of ill effects of undercrowding.

To be sure, overcrowding always produces ill effects, and these can always be demonstrated at some population density. On the other hand, the ill effects of undercrowding cannot always be shown, though frequently they can. In generalized curves the matter may be summarized thus: Under certain conditions[107] we find the curve running like the diagram in Figure 2A, when height above base line gives the strength of the biological action being measured, and distance to the right shows a steadily increasing population density. Under these conditions only the ill effects of overcrowding are visible, and the optimum population is the lowest possible. This is the modern expression of what used to be called the struggle for existence. In the more poetic post-Darwinian days this struggle was thought of as so intense and so personal that an improved fork in a bristle or a sharper claw or an oilier feather might turn the balance toward the favored animal. Now we find the struggle for existence mainly a matter of populations, measured in the long run only, and then by slight shifts in the ratio of births to deaths.

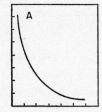 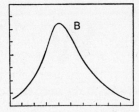

FIGURE 2. *A. Under some conditions the strength of biological action which is being measured is greatest with the smallest population, and decreases as the numbers increase. B. Under other conditions there is a distinct decrease in the strength of the measured biological reaction with undercrowding (to the left) as well as overcrowding (to the right).*

A second type of phenomena is represented by a curve with a hump near the middle, as shown in Figure 2B. Again, height above the base line measures the strength of some essential biological process or processes, such as longevity; distance to the right gives increasing population densities. The harmful effects of overcrowding, indicated by the long slope to the right, are still plainly evident, but there is also apparent a set of ill effects associated with undercrowding which are shown by the downward slope to the left. Many have written pointedly about overcrowding, and, although there is still much to be learned in that field, it is in the recently demonstrated existence of the mechanisms and implications of undercrowding that freshness lies. Without for one minute forgetting or minimizing the importance of the right-hand limb of the last curve, it is for the more romantic left-hand slope that I ask your attention.

Perhaps the simplest and most direct demonstration of certain harmful effects of undercrowding comes from an experiment which I understand is carried on spontaneously among undergraduate men at certain universities and colleges, of which X, or perhaps better, Y, is an example. A certain number of men gather together in a limited space under artificial light and undertake to consume a more or less limited amount of stronger or weaker alcohol. If there are many men present in proportion to the amount of alcohol, relatively little or no harm will result from the experiment. If there are very few men and much alcohol, there may be garage bills and other important repairs to be made.

In one way or another similar tests have been carried out in the laboratory with a variety of poisons and many kinds of animals. Again I choose from the mass of available evidence the results of a simple and clean-cut experiment to illustrate the same point with nonhuman animals.

Everyone is acquainted with goldfish; they are hardy forms or they would not be alive today in so many goldfish bowls. Colloidal silver, in the commercial form of argyrol is also well known. Colloidal silver, that is, the finely divided and dispersed suspension of metallic silver, is highly toxic to living things, including even the hardy goldfish.

In the experiment in our laboratory[8] we exposed a set

of ten goldfish in one liter of diluted colloidal silver, and at the same time placed each of ten similar goldfish in a whole liter of the same strength of the same suspension. This was repeated until we had killed seven lots of ten goldfish and their seventy accompanying but isolated fellows. Without exception, the mean survival time of the group exceeded that of the accompanying isolated fishes. The mean survival time for the seven sets of isolated fishes was 182 minutes; for the seven accompanying groups of ten each, 507 minutes. These values are summarized diagrammatically in the following table.

Survival of goldfish in colloidal silver, in minutes

NUMBER GROUPED	NUMBER ISOLATED	DIFFERENCE	STATISTICAL PROBABILITY
7 × 10	70 × 1		
507 min.	182 min.	325 min.	P < 0.001

Any biological experiment has a large number of so-called variables; that is, some factors are difficult or impossible to bring under complete control. Hence it is customary to make paired experiments if possible, in which one set of conditions (those of the group, in this instance) will differ from another lot (those of the isolated goldfish) only by the one difference, in this case of grouping and isolation. The results with these fish can then be analyzed by statistical methods to find the probability of getting like results merely "by chance." These methods are now so simple that even I can make the calculations. They are as accepted a technique as is the paired experiment.

With the goldfish there is less than one chance in a thousand of getting as great an average difference with the same number of random samplings. This is the same as saying that probability, or P, for short, is less than 0.001. Students of statistics have found that when P=

0.05 or less—that is, when there are fewer than five chances in a hundred of such a thing happening as a result of random sampling or "chance"—the results may well be significant as statistics go. The chances of observed differences being significant are better the lower the probability of their being duplicated by accident or by chance; technically speaking, this means that the smaller the fraction P is said to equal, the greater the probability of real significance in the results under consideration.

We make such tests of our experimental results continually, to find how we are getting on, and I shall give probabilities repeatedly. In doing so it must be remembered that these test the data, not the theory—and that the data may vary significantly for unknown reasons, even when we think we are in full control of the situation; and that because there is only one chance in one hundred, or ten thousand, or a million that a thing may happen by "chance" does not mean that it will never happen through what we call an accident; merely that the chances of its happening so, our evidence being what it is, are on the order of one in one hundred, or ten thousand, or a million.

I will digress even further into the realm of coincidence. A Negro friend of mine spent a summer in Europe and while in Paris visited the art galleries of the Louvre. While there he saw a Negro woman busy looking at pictures and on coming closer discovered that she was his own aunt. Neither had any idea that the other was in Europe. With no prearrangement, what is the probability that an American Negro from Chicago will meet his aunt in the Louvre? Yet it did happen this once without in any way shaking the probability principle.

Perhaps the digression is not so great as might appear at first glance, for we need some common understanding of the practical working of statistical probability; all of modern science, the more as well as the less exact, is built on it.

To get back to our goldfish: those in the groups of ten lived decidedly longer than their fellows exposed singly to the same amount of the same poison; and significantly so. But why? Others had made that experiment with smaller animals, and had decided that the group gave off a mutually protective secretion that would protect the particular species and none other. One reason we were working with goldfish was that they are large enough so that we could use approved methods of chemical analysis in finding where the silver went. The balance sheet from such tests showed that we could account for all the silver present. With the suspensions which had held ten fish, much of the silver was precipitated, whereas in the beakers that had held but one fish almost all the silver was still suspended.

When exposed to the toxic colloidal silver the grouped fish shared between them a dose easily fatal for any one of them; the slime they secreted changed much of the silver into a less toxic form. In the experiment as set up the suspension was somewhat too strong for any to survive; with a weaker suspension some or all of the grouped animals would have lived; as it was, the group gained for its members a longer life. In nature, they could have had that many more minutes for rain to have diluted the poison or some other disturbance to have cleared it up and given the fish a chance for complete recovery.

With other poisons, other mechanisms become effective in supplying group protection. Grouped *Daphnia*,[56] the active water fleas known to all amateur fish culturists, survive longer in overalkaline solutions than do daphnids each isolated into the same volume of a similar solution. The reason here is simple. The grouped animals give off more carbon dioxide, and this neutralizes the alkali. Long before the isolated individual can accomplish this, it is dead; in the group those on the outside may succumb, though if the number present is large enough even they may be able to live until the environment is brought under temporary control.

Frequently the protective mechanism is much more complex. With many aquatic animals, other things being equal, isolated animals consume more oxygen than if two or more share the same amount of liquid. By one device or another, grouping frequently decreases the rate of respiration of recently aggregated animals. Several of these devices are known to us. Professor C. M. Child showed many years ago[35] that when animals are exposed to a strongly toxic material, which kills within a short time, those with the higher rate of respiration, though otherwise similar, die first. This has been applied to group biology by direct tests, and it has been shown that the group, by decreasing the rate of oxygen consumption of its members, makes them more resistant to the action of relatively strong concentrations of toxic materials.

The situation is more complex. Exposed to relatively weak, slow-acting poisons, animals adjust better and live longer if they carry on oxidative processes at a relatively high rate. Interestingly enough, after the initial quieting effect of being aggregated, unfed groups of many kinds of animals have a higher rate of oxygen use than do the same number of isolated individuals given otherwise similar treatment. Hence, when rate of oxygen consumption is important, such groups give protection both for strongly toxic quick-acting poisons and for those whose slower action gives time for the animals to become adjusted to the toxic conditions imposed upon them.

Perhaps I have said enough to show that under a variety of conditions groups of animals may be able to

live when isolated individuals would be killed or at least
more severly injured by unaccustomed toxic chemical elements, strange to their normal environment.

Will the same relationship hold in the presence of
changes in *physical* conditions? There is a considerable
and growing lot of evidence that massed animals, even
those that can be called cold-blooded, are harder to kill
by temperature changes than are similar forms when they
are isolated.[57], [138] This interests us because massing of
such animals at the onset of hibernation was recognized
as one of the early exceptions to the rule, now out-
grown, that crowding is always harmful.

The exploration of temperature relations is a time-honored field. I prefer to take up a newer though related area,
that of the effects of ultra-violet radiation. In 1937 Miss
Janet Wilder and I began exposing a common planarian
worm to ultra-violet radiation, to find whether there
was any group protection from the well-described lethal
effect of ultra-violet light on these worms.[18]

In lots of twenty, worms of similar size and the same
history were placed together in a petri dish and exposed
to the action of the ultra-violet light long enough so
that they would disintegrate within the next twelve hours.
Half of them, that is, ten worms, were then placed together in five cubic centimeters of water and each of
the other ten was put into five cubic centimeters of similar water. Grouped and isolated worms were treated alike
in every way, except that after irradiation together,
half were grouped and half were isolated.

For one purpose or another we have repeated this simple experiment a great many times with a variety of waters, and with experimental conditions adequately controlled. Some of the things we have found out are:

If the worms are crowded under the ultra-violet lamp
so that they shade each other, the shaded ones are defi-

nitely protected. When such crowding is eliminated and by constant watching and stirring, if needed, the worms are kept approximately equally spaced during exposure, even then the grouped worms survive longer than the isolated. Some of the relationships are shown in Figure 3.

Each block represents the survival time of several series of worms. The figures at the top of the block give the average length of survival in minutes. The blocks are constructed so that the worms surviving longer, which in

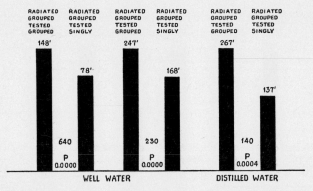

FIGURE 3. *Planarian worms that have been exposed to ultra-violet radiation disintegrate more rapidly if isolated than if grouped.*

each case are the grouped worms, are given as 100 per cent, regardless of the time taken; while the isolated worms, which had been irradiated in the same dishes as their accompanying groups, survived on an average of 78 per cent and 77 per cent respectively in the two tests with well water, and only 61 per cent in the test in distilled water. The numbers between the blocks show the number of worms averaged for each block; that is, the number of pairs of worms for which results are summarized. The statistical significance given in terms of P is very high in each case.

The number present during exposure is important, as well as the number present during the time when it is being determined how long the animals will survive. Such data are summarized in Figure 4, which is built on exactly the same principle as Figure 3. Worms radiated when crowded (left-hand block) and then tested when isolated, survived 517 minutes, whereas accompanying

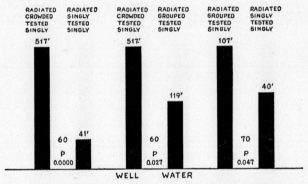

FIGURE 4. *Planarian worms survive exposure to ultra-violet radiation better if much crowded while being radiated, or even partially crowded, even though all are isolated after a few minutes of irradiation.*

worms which had been radiated singly as well as tested when isolated, lived only 24 per cent as long. Those radiated in a group and tested singly (middle block) lived 55 per cent as long as those which had been radiated in a crowd and then were isolated to observe the effects of radiation. It will be remembered that these crowded worms actually shaded each other and so gave physical protection from the ill effects of ultra-violet light. Finally (on the extreme right) is a diagram of the fact that worms radiated and tested singly lived only 62 per cent as long as those radiated in groups of 20 worms and then tested singly. Again the figures between the blocks give the number of pairs tested. P, the statistical probability, shows that all these data must be taken seriously even though there is decreasing significance as the percentage of difference of average survival time decreases.

In the two cases just outlined, mass protection has been demonstrated, first against the presence of toxic materials, and secondly against the ill effects of exposure to lethal ultra-violet rays. To complete the picture I have now to describe the results of exposing animals to harmful conditions in which the difficulty is caused by the absence of elements normally present in their natural environment. The experiment has been made on aquatic animals in a number of ways—for example, by putting fresh-water animals into distilled water; but it is easier to demonstrate when marine animals are placed in fresh water.

Again I select one experimental case from several available. Near Woods Hole, on Cape Cod, a small flatworm *Procerodes* (Figure 5) lives in certain restricted areas in large numbers. They are most abundant along a stony stretch at about the low tidemark or a little beyond it.[4] There, if one finds the proper location, one may take from ten to fifty flatworms from the lower surface of a single small stone. Usually they are more or less clumped together. They are not easy to see, since each is only a few millimeters long and all are of a dull gray color. Once seen, they are hard to detach, for the posterior end has a muscular sucker, by means of which the animal can cling pretty securely even to smooth stones. When these worms are put into fresh water—pond water, for example—they swell greatly and soon begin to disintegrate.

If these flatworms are washed thoroughly to remove sea water from their surfaces, and then placed in fresh water, a certain proportion of the grouped animals survive decidedly longer than isolated worms. The first worms to

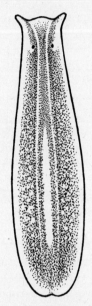

FIGURE 5. *The small marine flatworm* Procerodes.

die in the group do so almost as soon as the first isolated worms. As the dead worm disintegrates it changes the surrounding water; we say it *conditions* it; and as a result of this conditioning the remaining worms of the group have a better chance of life.

For more careful experimentation, a sort of worm soup was prepared by killing a number of well-washed worms and allowing them to remain in the water in which they had died and so condition it. Freshly collected *Procerodes* lived longer in such conditioned water than their fellows that were isolated into uncontaminated, clean pond water. The difference between the two waters was only that caused by the fact that in one the worms had died and disintegrated, whereas the other was clean. This difference

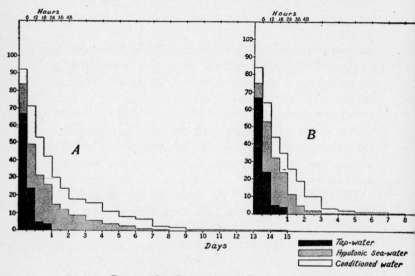

FIGURE 6. *Procerodes die more rapidly if transferred to pure fresh water than in dilute sea water, but live longer if placed in fresh water in which other* Procerodes *worms have died, even though the total amount of salt is the same as in the dilute sea water.*

in survival persisted even when, to make the test more revealing, the total amount of salt in the two waters was made identical by adding some dilute sea water to the clean pond water. Results from these experiments are shown in Figure 6. In this chart, distance above the base line gives the percentage of survival, and distance to the right shows time of exposure. It will be noted that the worms lived decidedly longer in the conditioned water than they did in dilute sea water of the same strength of salts.

The mechanism of this superficially mysterious group protection is now known.[97] The dead and disintegrating worms, or, more slowly, the living worms, give off calcium into the surrounding water, and calcium has a protective action for marine animals placed in fresh water or for fresh water animals put into distilled water, a protective action which is out of all proportion to its effect in increasing the osmotic pressure of the water. We can demonstrate that this is in fact the mechanism of such group protection. For example, we can analyze the water which worms have conditioned, find the amount of calcium that has been added, and by adding that amount directly get the same results that we do from the conditioned water.

This explanation is not yet complete—no scientific explanation ever is—but we have demonstrated that what was for a time a very mysterious group protection is in fact in this case an expression of calcium physiology. The further developments on the subject await exact information concerning the details of the physiological effects of calcium.

It is probably of more direct human interest to know that under many conditions bacteria will not grow if only a few are inoculated into an animal—man, for example; whereas with a larger inoculation they may grow abundantly.[37] Gentian violet is a poison for many bacteria and in regular medical use for that purpose. In one well-studied case (Figure 7) bacteria belonging to the species *Escherichia coli* failed to grow on agar containing gentian violet, if singly inoculated on it; only when thirty or more bacteria were inoculated did steady and regular growth occur. With the goldfish spoken of earlier, the mass protection was largely or wholly inoperative when the group of ten was exposed to ten times the amount of toxic colloidal silver to which a single fish was exposed. With these bacteria, however, such quantitative limitations did not hold; thirty organisms were found to make harmless at least two hundred times the amount of poison normally neutralized by an isolated

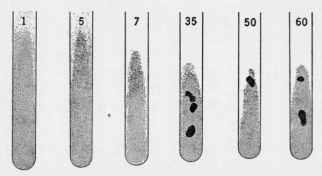

FIGURE 7. *Bacteria frequently do not grow if inoculated in small numbers; here different numbers of* Escherichia *were inoculated into a medium containing gentian violet.*

bacterium. This difference between the change which thirty bacteria can effect together as compared with what they can accomplish if isolated has been called an expression of the communal activity of bacteria. There is a fairly large and growing literature on this subject indicating that when only one or a few bacteria, even if strongly pathogenic, gain access to the human body that they are likely to be killed by various devices that aid in resisting infection. It is fortunate for their victims that bacterial infections normally tend not to take unless the inoculum is somewhat sizable or unless a smaller dose is frequently repeated.

Mass protection is known to occur among spermatozoa. Many animals, especially those that live in the ocean, shed their eggs and spermatozoa into the sea water, and fertilization takes place in that medium. Dilute suspensions of such spermatozoa lose their ability to fertilize eggs much sooner than if they are present in greater concentration. It is routine laboratory practice in experimenting with such animals as the common sea urchin, *Arbacia*, to keep sperm in a cool place, densely massed outside the body, for some hours. Small drops can be withdrawn as needed for experimentation, greatly diluted, and used almost immediately to fertilize eggs. When such dilute suspensions have long since lost their fertilizing power the sperm in the original dense mass are still potentially as active as ever.

So far we have been considering mass effects, the survival value of which, if any, was shown by increased length of life, often under adverse circumstances. Under many different conditions and for a variety of organisms, the presence of numbers of forms relatively near each other confers protection on a part of those grouped together or even on all present.

It is possible to go a step farther and demonstrate a more actively positive effect of numbers of organisms upon each other when they are collected together. Again I select a case for close scrutiny: that of crowding upon the rate of development in sea-urchin eggs.

Arbacia, mentioned above, is the common sea urchin of coastal waters south of Cape Cod (Figure 8). It has been much used in studies of various aspects of development, particularly by the biologists who gather each sum-

mer in the research laboratories at Woods Hole, Massachusetts. There are several reasons for its popularity.
These urchins are abundant in near-by waters and are
readily mopped up by the tubful. They can be kept in
good condition for some days in the float cages, and eggs
and sperm are readily procured as needed. Also the breeding season of *Arbacia* extends through July and August,
which are favored months for research at the seaside.

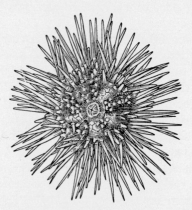

FIGURE 8.   Arbacia, *the common sea urchin of southern New
England, shown from the upper surface.*

For years biologists at Woods Hole have studied the
embryology and physiology of developing sea-urchin
eggs. They have built up a painstaking, almost a ritualistic,
technique for handling glassware, towels, and instru-
ments. The procedures require as rigid a cleanliness as does
a surgical operation. Consequently it was not surprising
when I first took up the study of *Arbacia* to have one of
my frankest friends among the long-time workers on their
development voice what was apparently a common feeling. He asked pointedly whether I thought I could come
into that well-worked field and without long training

find something he and his associates had overlooked. Such frank skepticism was refreshingly stimulating and added to the normal zest of biological prospecting.

The shed eggs of *Arbacia* are about the size of pin points and are just visible to the naked eye. The spermatozoa are tiny things; the individual sperm are invisible without a microscope although readily seen when massed in large numbers. When a few drops of dilute sperm suspension are added to well-washed eggs, one spermatozoan unites with one egg.

After some fifty minutes at usual temperatures, the egg divides into two cells. We call this the first cleavage. Thirty or forty minutes later a second cleavage takes place and thereafter cleavages occur rapidly. Within a day, if all goes well, such an egg will have developed into a freely swimming larva. Other things being equal,[12] the time after fertilization to first, second, and third cleavage is speeded up for the crowded eggs. Typical results and some of the methods are shown in Figure 9. With appropriate experimental precautions, some eighteen hundred eggs were introduced into a tiny drop of sea water. Near by on the same slide forty similar eggs were placed in a similar drop and the two were connected by a narrow strait as shown in the figure. A few eggs from the larger mass spilled over into this strait. The whole slide was placed in a moist chamber to avoid drying and was examined from time to time. In a trifle over fifty-five minutes, half the eggs in the densest drop had passed first cleavage. A half-minute later, 50 per cent of those in the strait were cleaved; and twenty seconds later, half of the more isolated ones had divided. The time to 50 per cent second cleavage ranged between eighty-four minutes for the crowded eggs and over eighty-six and a half minutes for the isolated ones.

This was repeated with four thousand eggs or there-

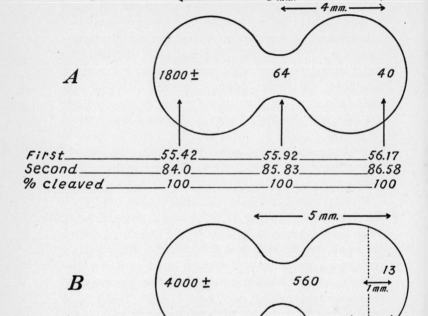

FIGURE 9. *Eggs of the sea urchin,* Arbacia, *cleave more rapidly in dense populations than if only a few are present. Figures below the diagrams give time in minutes.*

abouts in the denser population, almost six hundred of which spilled through and formed a flat apron over the bottom of the second drop, in which there were thirteen other eggs scattered singly about the relatively unoccupied space. Under these conditions the time to 50 per cent first cleavage was approximately fifty-two, fifty-eight, and sixty minutes respectively, and the difference at the middle of the second cleavage was even greater.

In association with Dr. Gertrude Evans, this experiment was repeated in many different ways, and there remains in my mind no doubt but that under a variety of conditions the denser clusters of these *Arbacia* eggs cleave more rapidly than associated but isolated fellows.

Under the conditions tested, the stimulating effect of crowding could be detected when sixty-five or more eggs were present in the more crowded drop and twenty-four or fewer eggs made up the accompanying sparse population.

Within twenty-four hours, under favorable conditions, the cultures are full of free-swimming *Arbacia* larvae with characteristic arms, which are known as plutei. When all our available data collected the first day after fertilization are compared, there is again no doubt that the more crowded cultures usually develop more rapidly than accompanying but sparser populations. However, it must be recorded that throughout the whole series there were occasional isolated eggs that developed as rapidly as the best of the accompanying denser populations. Such eggs and embryos were exceptional in our experience; the fact that they exist indicates clearly that under the conditions of our experiments crowding, although usually stimulating, was not absolutely necessary for rapid cleavage and early growth.

In this connection it is interesting to note that others have prepared an extract from sea-urchin eggs and larvae that is growth-promoting[102] and one that is growth-inhibiting. As has also been found with goldfish, the growth-accelerating principle seems to be associated with the protein fraction of the extract. When the whole extract is used, it is growth-inhibiting and produces the same results as overcrowding. The point I have made is that with the sea-urchin eggs, under the conditions of our experiments, there is also an ill effect of undercrowding, and that there is an optimum population size for speedy development, which is neither too crowded nor too scattered.

Much similar work has been done with the effects of numbers on the rate of multiplication with various protozoans. Again I shall have to select results from the mass of available evidence. The late T. Brailsford Robertson,[117] of Australia, announced back in 1921 that when two protozoans of a certain species were placed together the rate of division was considerably more than double that which resulted with only one present. It should be noted that during the time of these experiments and in all these protozoa that we are considering reproduction was entirely asexual, by self-division of the original animal. I subjected the data in Robertson's original paper to statistical analysis and found that there were only thirteen chances in a thousand of getting as great a difference by random sampling. Such results must be taken seriously (Figure 10).

They were. And the period after 1921 was enlivened for some of us by denials from one first-class laboratory after another that there was anything significant in Robertson's data. Robertson himself rechecked and confirmed his results, though his explanations of them tended to vary. For the moment we are not concerned with the explanations; but what are the facts? The first extensive corroboration from outside Robertson's own laboratory came from the work of Dr. Walburga Petersen at Chicago. When she cultured the common *Paramecium* in small volumes of liquid, she obtained the same results as had Robertson's critics, but when she used relatively larger volumes of the same culture medium, a cubic centimeter, more or less, she got an increase in division rate with the presence of a second individual, as Robertson had found it in the Australian form he had studied.

Still the critics were not convinced. Accordingly Dr. Willis Johnson repeated this whole study using a different protozoan, one of the *Oxytricha*.[76] When sister cells from pure-line cultures were used there was no difference at the end of the first day, whether the *Oxytricha* were in-

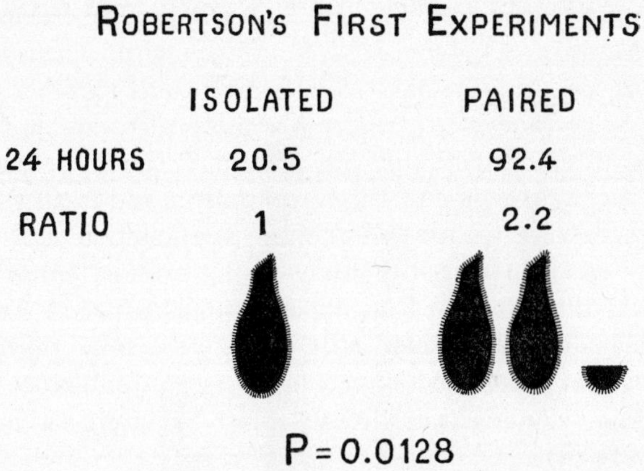

ROBERTSON'S FIRST EXPERIMENTS

	ISOLATED	PAIRED
24 HOURS	20.5	92.4
RATIO	1	2.2

$$P = 0.0128$$

FIGURE 10. *Robertson found that when two protozoans were placed together each yielded over twice as many as when the same number of similar protozoans were isolated.*

troduced singly or in pairs into one or two drops of good medium. Later, the cultures started with one organism always were ahead. With larger volumes, two organisms showed a higher rate of reproduction per original animal at the end of the first day than if started with a single protozoan.

Again Robertson's results were confirmed for larger volumes and those of his critics for smaller volumes. But Johnson had only started. He knew from the work of

52

others that if a protozoan is washed through several baths of sterile water the associated bacteria are rinsed off. Then if the washed protozoan is put into a weak solution of the proper salts, into which has been introduced known numbers of the bacteria on which they normally feed, the problem can be studied with a controlled food supply, both as to kind and amount.

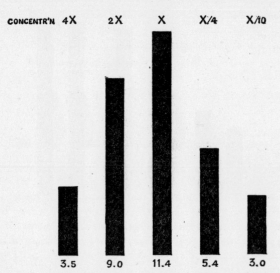

NUMBERS OBTAINED IN 24 HOURS FROM THE ISOLATION OF OXYTRICHA INTO CONSTANT VOLUMES WITH DIFFERENT CONCENTRATIONS OF BACTERIA

CONCENTR'N 4X   2X   X   X/4   X/10

3.5   9.0   11.4   5.4   3.0

FIGURE 11. *The ciliate protozoan* Oxytricha *reproduces more rapidly with a certain limited number of bacteria present than with either more or fewer. (From Johnson.)*

This he proceeded to do. He found a common bacterium on which his sterile *Oxytricha* would grow and reproduce faster than in the ordinary medium. He made standard suspensions of these bacteria in sterile salt solution, at what we may call an X concentration. The bacteria could reproduce little, if at all, in the salt medium, so that he knew how much and what kind of fodder he was feeding his washed protozoans.

The results of varying the amount of food are shown in Figure 11. With X concentration, in twenty-four hours one animal produced about eleven progeny. With 2X

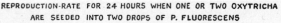

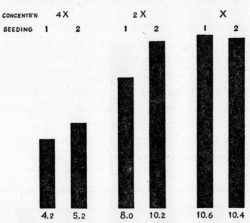

FIGURE 12. *In the denser suspensions of bacteria, the protozoans divide more rapidly when cultures are inoculated with two protozoans than if started with a single individual. (From Johnson.)*

concentration, isolated sister cells produced nine, and with a 4X concentration other isolated sister cells produced but three and a half. The rate of reproduction also decreased when less than X bacteria were present.

Now he was ready for the grand Robertson test, except that by this time nearly all the factors were controlled.

The results are shown in Figure 12. With X concentration it made no difference whether he started his small cultures with one or with two sterile animals. With 2X concentration, the cultures started with two individuals did as well as in X concentration, but those which were started with only one individual lagged definitely, producing only 80 per cent as many animals in twenty-four hours. With 4X concentration even the culture started with two *Oxytricha* was slowed down, but not so much as that started with only one. He had shown that, in the presence of an excess number of bacteria, cultures seeded with more than one bacterium-eating protozoan thrive better than if but one is introduced. Not content with this Johnson took another species and tried it all over again with the same results.

From all this careful work we judge that the facts on this particular aspect of the effects of numbers present on the rate of asexual reproduction seem now to be straight; but what about their explanation? This, as it turns out, also interests us. Robertson advanced the following hypothesis to explain the results which he had observed. During division each nucleus retains as much as possible of an essential, growth-producing substance with which it was provided, and adds to it during the course of growth between divisions. At each division, however, this substance is necessarily shared with the surrounding medium in a proportion that is determined by its relative solubility in the culture water and by its affinity for chemical substances within the nucleus. The mutual speeding of division by neighboring cells results from each cell's losing less of this necessary substance because of the presence of the other. The more of this growth-promoting substance there was in the cell, Robertson thought, the faster would be the division rate; so that any circumstance which would conserve the limited supply would tend to speed up processes leading to cell division.

Stripped to essentials, this hypothesis says that as a result of the presence of a second organism both lose less of an unknown something which is essential for divison than would happen if but one were present. Returning to the problem after the criticisms of half a dozen years, Robertson affirmed that all the data and conclusions on the subject that had been issued from his laboratory remained valid save that they might apply to the associated food organisms and not to the protozoans themselves.

Johnson has paid considerable attention to this problem, and has concluded that the results which he has observed can be explained as being caused by bacterial crowding; that the larger number of protozoans introduced into dense cultures thrive best because they are able to reduce the bacteria to density optimal to the protozoa faster than can their isolated sister cells; and therefore they show a higher rate of reproduction.

This does not seem to be the whole story; for from points as distant as Baltimore[89] and Jerusalem[111] I have reports from trustworthy men that with still simpler protozoans they are getting results which suggest that some modification of Robertson's hypothesis may be correct after all. These organisms stimulate each other to more rapid growth merely by their presence in the same small space.

With fine courtesy, the late Professor S. O. Mast, of Johns Hopkins, placed a report of his experiments in my hands in advance of publication and permitted me to summarize his results. He finds that populations of a flagellate protozoan grow more rapidly in a sterile medium of relatively simple salts when larger numbers are introduced than if the cultures are started with only a few organisms.

In the decade and more since the present summary was first written, much confirming evidence has been published. As expected, it has been shown that complicated problems such as those dealing with the rate of population growth are controlled by more than one mechanism.

The suggestions from the simpler protozoans, taken together with other aspects of the mass physiology of protozoa, which have been only partially reviewed here, and with the acceleration of development demonstrated for sea-urchin eggs, encourage me to renew a suggestion made some years ago,[3] which has, so far as I am aware, been overlooked to date.

Let us go back to consider the case of external fertilization among aquatic animals. When spermatozoa and eggs are shed into sea water by sea urchins or other marine animals, their length of life is distinctly limited. If a sperm fails to make contact with an egg during the fertilizable period, death results, probably from starvation for the spermatozoa, perhaps from suffocation for the egg. This means that the animals of the two sexes must be fairly close together if there is to be a union of the shed sexual products. The most vigorous sperm of the sea urchin *Arbacia* can travel in still water about thirty centimeters—that is, about one foot and two inches.[62] Spermatozoa of these animals diluted a few thousand times can survive from three to twelve hours; the majority last no longer than seven hours. If a current catches it, such a sperm can travel many times thirty centimeters, but even in sea water the sexes must be relatively aggregated if fertilization is to be successful. In fresh water, the life of shed gametes is much shorter. After ten minutes, eggs of the pike lose the power to be fertilized,[112] and the longevity of sperm of certain freshwater fishes is less than a minute, so that in fresh water aggregation is even more essential. With animals that re-

quire internal impregnation the necessity for close cooperation between at least two individuals is obvious. Such
considerations must be fundamental for the long-recog-
nized breeding aggregations of animals, especially of
those that shed eggs and sperm into surrounding water.

Mass relationships may be even more important sexually, and here I come to the new suggestion: perhaps they
had a hand in shaping sex itself. Presumably sexual evolution started, as may be found today in plants, with a time
when all gametes of any one species were similar. Under
these conditions a first step toward the union of two reproductive elements could be supplied by the greater
well-being fostered by the presence of more than one
gamete within a limited area, as even the simpler protozoan are stimulated to non-sexual division today by the
near-by presence of another of the same species. In the
survival value existing for separate living cells before actual sexual union took place we can find a logical beginning for the action of selection, which would in turn,
with present known values, result in the establishment
of the sexual phenomena as they appear today. These
fields have not been sufficiently explored to allow for
more than this flash of imagination, which future researches may verify or discard.

At this point it would be well to pause and look back
over the road we have traveled thus far. The charts[5]
shown as Figures 13A and B show that most of our evidence has come from fairly well down among the simpler
forms of life. I have called attention to mass protection
of one sort or another among bacteria, planarian worms,
goldfish, and the simpler crustaceans. Actually there are
in scientific literature good cases of mass protection for almost all the animals shown in these charts; and where
exact information was lacking, as for example among the

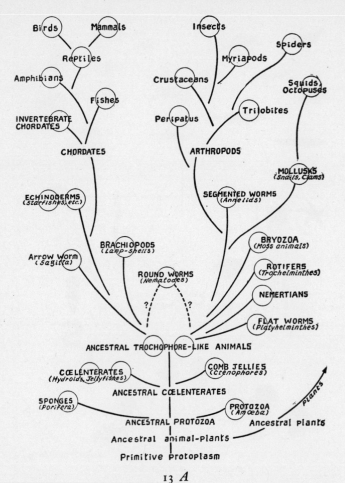

13 *A*

FIGURE 13. *A suggestion concerning the ancestral relations within the animal kingdom. The circles in A and B allow cross-identification. (From Allee in* The World and Man.)

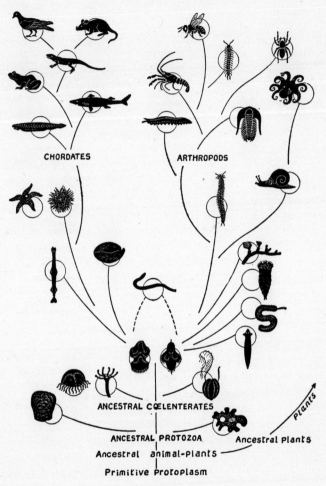

CHORDATES   ARTHROPODS

ANCESTRAL CŒLENTERATES

ANCESTRAL PROTOZOA   Ancestral Plants

Ancestral   animal-plants

Primitive Protoplasm

Plants

13 B

rotifers, this was a result only of lack of interest in conducting experiments on this point with these animals. Eventually it turned as predicted, that mass protection for rotifers could easily be demonstrated.[17]

I have also shown active acceleration of fundamental biological processes as a result of numbers present for sea-urchin eggs and larvae, and for various protozoans. These have been given in some detail, which has not left time for similar demonstrations among regenerating cells of sponges; nor have I time to tell how hydra have been saved from depression periods by the use of self-conditioned water. I have mentioned but not elaborated the fact that grouped animals frequently have different rates of respiration as compared with their isolated fellows. This has been recorded widely in the animal kingdom, notably among planarians, certain lower crustaceans, some starfish, fishes, and lizards, and for some, at least, associated survival values have been demonstrated. To this extent, then, I have given the crucial evidence I promised earlier that a sort of unconscious proto-cooperation or automatic mutualism extends far down among the simpler plants and animals.*

These charts should illustrate one other point. The insects stand at the apex of one long line of evolution; mammals and birds are at the peak of another line of evolution; the two have been distinct for a very long time. This view of evolution indicates that the ancestral tree of animals is not like that of a pine tree with man at the very top and insects and all the other animals arranged as side shoots from one main stem. Rather, there are at least two main branches that start low, as in a well-pruned peach tree. Both rise to approximately equal heights, in-

* A. Montagu in *On Being Human* published in 1950 reviews much of this evidence.[93]

dicating correctly that in their way the insects are as specialized as the birds or mammals. Since both insects and mammals have developed closely knit social groups, this is further evidence that there is a widely distributed potentiality of social life. We shall return to this subject later.

Aggregations of Higher Animals.

Chapter V

A great deal of skepticism is necessary in science, if progress is to be even relatively steady and sound. Not only must the scientist be skeptical of advance reports of new results until he has seen the supporting evidence, no matter how stimulating the thesis and how well it would explain material already gathered, but in fields that lie near his own researches it is necessary to bring the problem into his own laboratory if possible and there examine the validity of the evidence itself. This repeating of experiments in order to check the first observer is sometimes also a testing of scientific courtesy, but every real scientist must be prepared to submit to it with the best grace possible.

It is demanded also that from time to time one should be skeptical of views long held, and of the evidence on which they were built up, particularly of the inclusiveness of the conclusions that have been drawn. Without my own fair share of this skepticism I should never have been drawn into what I knew from the beginning would be a long and laborious series of experiments concerning the effects of numbers present upon growth.

As long ago as the eighteen-fifties, Jabez Hogg,[70] an Englishman, found by experimenting that crowding decreased the rate of growth of snails and produced stunted adults. From that day to this there has been almost no break in the reported evidence that overcrowding reduces growth; the number of reports that crowding in any degree increases growth are relatively few.

This phenomenon has, however, been observed by enough workers using animals widely distributed through the animal kingdom to show that the retarding effect of undercrowding on growth is real. Before considering the implications of this statement let me review briefly some of the evidence.[3] Here as elsewhere I shall make no attempt to catalogue all the available evidence; the list would be impressively long but tedious.

It is relatively easy to show that mixed populations of many animals grow faster than if the same number of some one species are cultured together. The common experience of aquarium enthusiasts that the presence of the snails in aquaria increases the rate of growth and well-being of their fishes is a case in point. Their rule-of-thumb experience has been fully verified by careful laboratory experiments. A more crucial test involves individuals of the same species: all snails, let us say, or all goldfish. Is there some optimum size of the population at which individuals grow most rapidly?

For years I have been studying different aspects of this problem with the aid of a succession of competent, critical research assistants and associates all of whom have independently obtained the basic results I am about to describe.[9, 16, 17, 86]

We have used goldfish for our experimental animals, because they are inexpensive, easy to obtain, hardy under laboratory conditions, and able to stand daily handling.

In order to have a consistently constant water we make up a synthetic pond water by dissolving in good distilled water salts of high chemical purity. Into such water goldfish about three inches long are placed in sufficient number so that they will give a conditioning coefficient of about twenty-five. Let me explain: this coefficient is obtained by multiplying the number of fish by their average length in millimeters and dividing by the number of liters of water in the containing vessel. Living in this water the fish condition it by giving off organic matter and carbon dioxide. They are left in the water for twenty-one hours or so, while a similar amount of the same water stands near by under exactly similar conditions except for the absence of fish.

At the end of this time the clean control water is siphoned into a number of clean jars, and a small measured goldfish is placed in each. At the same time the conditioned water is siphoned, either with or without removing particles (that is, of excrement, etc.) that may be present, into similarly clean jars. A set of small measured goldfish, like those used in the control jars, are transferred into the conditioned water. These small "assay" fish have been feeding for about two hours before being transferred; the larger conditioning fish are allowed to feed for a somewhat longer time before being washed carefully to remove food residues and replaced in another lot of water to condition that.

Meantime the jars, 120 of them, are all washed carefully; and after this is done the experimenter has nothing more to do until the next day, except to put the laboratory in order, keep the temperamental steam-distilling apparatus running, test the water chemically in several ways, keep his records in order, and otherwise see that nothing untoward happens to make him or anyone else question the results.

After some twenty, twenty-five, or thirty days of such care, in which Sundays are included, again each fish is photographed to scale, as they were also photographed at the beginning of the experiment; the photographs are measured and the relative growth determined for the fish that have daily been placed into perfectly clean synthetic pond water, as compared with those which daily have been put into conditioned water, that is, into the water in which other goldfish have lived for a day.

During the course of an analysis of this problem we have performed this simple basic experiment many times. The first forty-two such tests, involving 886 fish, showed on the average about two units more growth for the fish in the conditioned, slightly contaminated water, than for those in clean water (Figure 14). These results have a statistical probability (P) of about one chance in a hundred million of being duplicated by random sampling. Hence we have demonstrated that under the conditions of our experiments the goldfish grow better in water in which other similar goldfish have lived than they do when they are daily transferred to perfectly clean water.

The problem that has been occupying us for some time is why this is so. What are the factors involved that make this slightly contaminated water better for young goldfish than a clean medium?

We have said that the conditioning fish are fed for two or more hours daily and are then washed off and placed in a fresh batch of water. Although the fish are never fed in the water they are conditioning, within a few hours after their transfer into it from the feeding aquarium the water becomes more or less cloudy with regurgitated food particles. These bits of food are taken in by the growth-assay fishes swallowing them along with the containing water.[13] When such particles are removed by filtering, the growth-promoting power of the conditioned water is greatly lessened, but it is not completely lost. In our experiments we found that suspended food particles accounted for 80 per cent or more of the increased growth in conditioned water over that in clean control water.

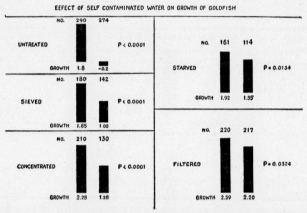

FIGURE 14. *Goldfish grow more rapidly if placed in various kinds of slightly contaminated (conditioned) water. The numbers above the columns show the number of fish tested. The longer column represents the growth in conditioned water.*

These experiments give certain suggestions concerning some other conditioning factors that may be acting. For example, we know that the skin glands of fish secrete slime (Figure 15). When we have made a chemical extract of this material we have frequently recovered a growth-promoting substance, apparently a protein, which was effective in stimulating growth when diluted 1 to 400,000, or even 1 to 800,000 times. At these dilutions it is not probable that this factor is affecting growth by furnishing food material.

There are, of course, other possibilities, many of which we have checked. The increase in growth does not result, for example, from a change in the total salt content of the water, for this does not change in our experiments; nor from differences in acidity or oxygen, nor, so far as

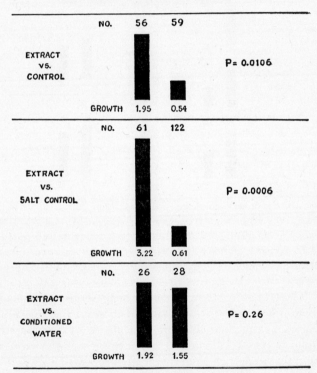

EFFECT OF PROTEIN EXTRACT FROM SKIN OF
GOLDFISHES ON GROWTH OF GOLDFISH

FIGURE 15. *An extract from the skin of goldfish frequently has growth-promoting power. The arrangement of the figure is on the same plan as was used in* FIGURE 14.

careful quantitative analyses have revealed, from changes in chemical elements present. We may be dealing with some sort of mass protection, such as was discussed in the last chapter, in which the conditioning fishes remove some harmful substance, but of this we have no real evidence.

Whatever the explanation, we are certain of the facts, and we know that we have demonstrated a device such that if in nature one or a few fish in a group find plenty of food, apparently without willing to do so they regurgitate some food particles that are taken by others, a sort of automatic sharing. Again, in water that changes from day to day, such stagnant-water fishes as goldfish, if present in numbers, are able to condition their environment, perhaps by the secretion of mucus, so that it becomes a more favorable place in which to live and grow.

Perhaps I have lingered too long over this one case; I am so close to the facts and to the tactics used in collecting them that they still seem interesting and important to me more than ten years after their first discovery. We have run the same experiment with positive results with a few other species of fishes; and we have also found by experimentation that certain fish will regenerate tails that have been cut off if several are present in the same water more rapidly than if each is isolated.[123] The same is true for the young tadpole-like larvae of salamanders, with which we have had experience. The explanation of the more rapid regeneration of such cut tails is probably relatively simple. The several animals together more readily bring the surrounding fresh water to approximately the salt content of the cut and regenerating tissues than can be done by a single animal placed in the same amount of water. This may not be the whole of the story but it is probably a significant part of it.

In both of these cases the additional growth of aquatic

animals, which occurs as a result of the presence of other animals of the same species, is produced in response to some sort of chemical which has been given off into the surrounding water. This may be nothing more than the unswallowing of surplus food by the conditioning fish. With animals whose tails have been freshly cut off the addition of salts to the water by the group may balance the osmotic tension at the cut surfaces and so favor regrowth. The exciting result of these studies lies in the suggestion that some less obvious growth-promoting substances may also be secreted into the surrounding water.

Animal aggregations frequently produce physical as well as chemical changes, and while we are considering the effect of numbers of animals present on the rate of growth of individuals it is interesting to examine one case in which growth-promotion appears to have been produced largely by changes in temperature. Such an effect has been reported more than once; it is most simply illustrated in a warm-blooded animal, this time the white mouse. The experiment was first performed in Poland, but the causal factors were then only partly recognized. It has been repeated in our laboratory, where significant steps have been taken toward its further analysis.

Vetulani, the original experimenter,[129] used closely inbred mice for his experimental animals. He measured the growth of males and females separately from the sixth and on through the twenty-second weeks of their lives. After rearrangement he followed them for ten weeks longer as a sort of control. Fresh food was supplied in abundance each day, and proper experimental conditions seem to have been maintained.

Growth during the first sixteen weeks of the experiment is shown in the accompanying graphs (Figure 16). All started off at approximately the same rate. After the fifth week of the experiment, however, it is clear that the isolated mice were growing most slowly, and they continued to do so as long as the experiment ran. The most rapid rate of growth was observed in those mice which were placed two to four per cage; those five to six

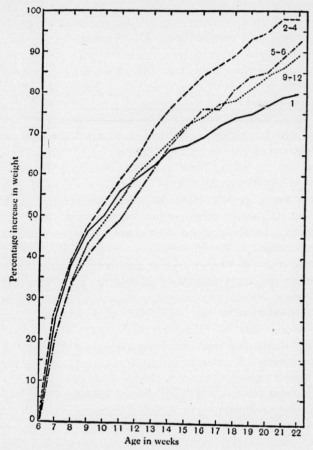

FIGURE 16. *White mice grow faster in small groups than in large ones; they grow slowest when isolated (solid line). (From Vetulani.)*

per cage grew next best, and only slightly below these came those living nine to twelve per cage.

Under the conditions of this experiment the isolated young mice were most handicapped, those most crowded were next, and those that were somewhat but not too crowded grew most rapidly. When the mice were re-arranged for a continuing period of ten weeks the same relations held, showing that it was the state of aggrega-tion rather than individual differences between mouse and mouse which was important in producing the differ-ences in growth rates.

Mr. E. Retzlaff,[115] the student who brought this work into our laboratory, tried first to repeat Vetulani's ex-periments in a room held at relatively high temperatures (29-30° C.). Under these conditions he found that inso-far as significant differences existed they showed that most rapid growth occurred with the isolated mice. When he lowered the room temperature to about 16° C., however, he obtained the same general effect as reported by Vetulani. It would seem, then, that in this case the opportunity to keep warm in a chilling temperature is one of the main factors in promoting growth of the crowded, but not too crowded, animals. This conclusion is strengthened by analyses of the temperature relations of mice, made by French physiologists,[34] which show that a mammal as small as a mouse has great difficulty in maintaining a constant temperature and rarely does so for extended periods of time. A change of external tempera-ture from 30° to 18° C. will cause a lowering of 0.4° in the body temperature of a resting mouse.

With such temperature lability it is easy to see that a few mice huddled together as is their habit could help each other to maintain their internal temperatures, con-serving energy for growth, whereas if isolated they must use much of their energy in keeping warm.

Vetulani observed another factor at work. Some of his mice had lesions of the skin which they treated by licking. When these were in the head region they could only be treated by another individual. Some of his isolated mice had such lesions when at the end of the first experimental period they were regrouped for further observation; these wounds were soon cured by their new nest mates.

When one turns from studying the rate of growth of individuals to that of populations of these higher sexual animals, many of the same principles can be observed working as were outlined in the last chapter for the growth of asexual populations of protozoans, in which overcrowding retards population growth and optimal crowding, at least in many instances, favors it.

With experimental populations of mice, for example, three long, laborious experiments made in Scotland[41] and in Chicago[116] have indicated that, under the conditions tried, the least crowded mice reproduce most rapidly. The same holds true for the well-studied fruit fly, *Drosophila*.[107]

Sometimes with these flies, but not with the mice, there is an indication of a more rapid rate of reproduction per female when more than the minimal pair is present. I have a strong suspicion, however, that one would get a more rapid rate of increase per number of animals involved if, instead of keeping the sexes equal in numbers, there were a ratio, let us say, of two females to one male.

We do know that with *Drosophila* the greatest numbers are produced when the feeding surface is relatively great but not too great.[68] These flies eat yeast; with too large a space, or in other words, with too few flies present, harmful wild yeasts or molds may grow too rapidly for the oversmall population of *Drosophila* to keep them under control. This suggestion is based on the assumption that with enough fruit flies present even these supposedly harmful plants would not produce marked effects.

Another well-studied laboratory animal, the flour beetle, *Tribolium*, under certain experimental conditions gives most rapid population growth at an intermediate population size rather than with too few or too many present. A study of data collected by Chapman showed that in a flour beetle's little world, a microcosm of thirty-two grams of flour, these beetles, during the early stages

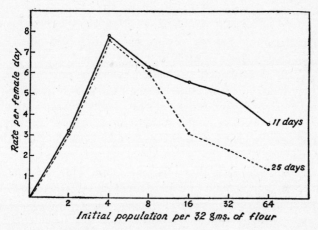

FIGURE 17. *Flour beetles reproduce more rapidly if more than one pair is present.*

of population growth, reproduce most rapidly per female with two pairs present (Figure 17). Reproduction is more rapid when four pairs or even sixteen pairs are present than if there is only one pair.[3]

This work of Dr. Chapman's was done for another purpose. We took it for an indication of possibilities, and Dr. Thomas Park looked into the matter independently.[97] He found the situation very much as it had originally appeared to be. A Scotsman named Maclagen had a curiosity along the same line and independently rechecked the whole matter with the same results.[87] Three separate workers in three different laboratories have now obtained essentially similar results with these same beetles, and the chances that all are mistaken are rather remote.

One of the three, Dr. Thomas Park, has proceeded to analyze the factors involved.[99, 100] He finds that the results come from the interaction of two opposing tendencies. In the first place, adult beetles roam at random through their floury universe. They eat the flour, but they may also eat their own eggs as they encounter these on their travels. This habit of egg-eating tends to reduce the rate of population growth, the more so the denser the population.

The second factor is the experimentally proven fact that up to a certain point copulation and successive re-copulation stimulate the female *Tribolium* beetles to lay more eggs, and eggs with a higher percentage of fertility. Thus the more dense the beetle population the more rapid its rate of increase. The interaction of these two opposing tendencies results in an intermediate optimal population in which more offspring are produced per adult animal than in either more or less dense populations.

It may be felt that I have been keeping too closely to the more or less artificial conditions found in the laboratory. It is true that in an attempt to bring the various aspects of the population problem under experimental control we have avoided those field observations which can only be recorded as more or less interesting anecdotes. We have now come to a point in our inquiry, however, at which it is necessary to move directly into the field.

Given the evidence at hand, that optimal numbers present in a given situation have certain positive survival values and some definitely stimulating effects on the growth of individuals and the increase of populations, we strike the problem of the optimal size of a population in nature. This is an exceedingly difficult question on which to obtain data. Suppose, therefore, that we simplify it by asking what minimal numbers are necessary if a species is to maintain itself in nature?

This inquiry is a direct attempt to find under natural conditions the application of the statement by the late Professor Raymond Pearl that "this whole matter of influence of density of population in all senses, upon biological phenomena, deserves a great deal more attention than it has had. The indications all are that it is the most important and significant element in the biological, as distinguished from the physical, environment of organisms."

Over and over again in the last decades I have asked field naturalists, students of birds, wild-life managers, anyone and everyone who might have had experience in that direction, how few members of a given species could maintain themselves in a given situation. Always until 1937 I found that, stripped of extra verbiage behind which they might hide their ignorance, the real answer was that they did not know.

And then I had two pieces of luck: I found a man and a scientific paper. My friend Professor Phillips, of South Africa, came to spend some weeks with us. He told us that the Knysna Forest, a protected woodland in South Africa, has an area of 225 square miles, fifteen miles on a side, and that this forest is the home of a herd of eleven elephants, which can also range outside the forest limits. On the other hand, the Addo Forest, of twenty-five to thirty square miles, supports a herd of twenty-four elephants.[108] Dr. Phillips thinks that the smaller herd is not maintaining itself, and that the larger one, under apparently less favorable conditions as regards available area of range, is at approximately the lower limit for keeping up its own numbers. He estimates that an elephant herd of about twenty-five individuals could maintain itself in an unrestricted range provided civilized man were absent.

He gave us a second example, of a herd of some three hundred springbok on a protected reserve of six thousand acres in the Transvaal, which was unable to maintain its numbers and became reduced to eighty or ninety, on its way toward total extinction.

It is well known that in the life of equatorial Africa the tsetse fly plays an important part. It carries the trypanosomes that cause the deadly disease "sleeping sickness" among man and his domestic animals, and that affect native game as well. The British colonial governments have been active in attempts to control the density of these fly populations. In general they are restricted to damp, low-lying forest. In districts where this is confined to the borders of watercourses, and hence where the fly belt has naturally a definite limit and is restricted in size, an ingenious fly trap has been used successfully. The trap takes advantage of the natural reactions of the tsetse fly. These are strongly positive to a slightly moving dark object a few feet above ground. With appropriate screening they can be caught as they fly toward such an object; they will fly up and fall back until they literally wear themselves out. It was at first thought that such a trap would be helpful chiefly in reducing the excess fly population; then, to the delight of the control officials, they found that when in these restricted fly belts the tsetse flies had been trapped down to a certain minimum population there was no need to catch the very last flies; below the minimum level those remaining disappeared spontaneously from the area. Nor did they return unless brought back in considerable numbers accompanying movements of game, or as a result of the slow extension of range from other infested areas. The work of the control officials in such regions thus was very much easier than had been anticipated.

Two pertinent cases concerning the minimum number below which a species cannot go with safety have come in part under my own observation. In 1913, my first summer at the Marine Biological Laboratory at Woods Hole, Massachusetts, the veteran scientists of the laboratory, at least those who still were willing to exhibit naturalistic enthusiasms, were greatly pleased at the visit of a flock of laughing gulls to the Eel Pond near the laboratory. The main breeding ground of these gulls is on Muskeget Island, off Nantucket. In 1850 the laughing gulls were abundant there; but they were exposed to the depredation of egg-takers and later, about 1876, to the attacks of men interested in obtaining their striking wings and other feathers to satisfy the millinery demand for feathers of native birds, which was then at its height.[55] Under this slaughter the colony was nearly wiped out; at its low point about 1880 there were not more than twelve pairs of laughing gulls left on Muskeget Island, and only a few of these bred. A warden was employed in a somewhat extralegal capacity by certain ornithologists who regretted seeing the species die out, and he was assisted by the captain of the local lifesaving crew in protecting the gulls from raids. Later changes in laws regarding protection of birds and the use of plumage in millinery gave more secure protection for the growing colony. For the first ten years the birds increased slowly, but thereafter more rapidly, until there are now thousands breeding on the island, and their range has spread to the mainland. In Woods Hole, at the present time, these birds whose return in 1913 excited so much comment are as common as the terns. In this case, a few breeding pairs,

78

nesting in a relatively safe place, were able to regenerate
the local population in less than fifty years; all that was
needed was protection from the predations of man.

The nesting colonies of gulls have attracted attention from many; a report by F. Fraser Darling concerning
certain relations between numbers of herring gulls in a
colony and breeding behavior, and survival of young gulls
on Priest Island off the northwest coast of Scotland is
suggestive.[44] There are indications that the members of
larger colonies stimulate each other to begin mating activities earlier than when the colonies are small and, what is
apparently more important, there tends to be a shorter
spread in the time from the laying of the first egg until
the last one is laid. This means that the breeding activities
are more intense while they last.

The period between hatching and the growth of the
first adult plumage is a crucial time in the life of young
gulls. While they are in the downy stage they are preyed
upon by outside predators; also at this time the gull
chicks that wander from their home nests may be pecked

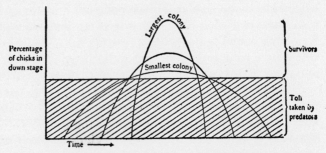

FIGURE 18. *The "spread" of time in which eggs are laid in
a colony of herring gulls affects the percentage that survive.
The smaller the colony, the longer the spread and the fewer
survivors. (From Darling[44] by permission of The Macmillan
Co.)*

to death by other members of the colony. The toll of the chicks is comparatively less, the shorter the time from the hatching of the first fuzzy young gull until the last one changes to a young fledgling with adult feathers. These relations are graphically shown in Figure 18.

Darling thinks that the greater success of the larger colonies lies not in any vague factor of mutual protection, but in the nearer approach to simultaneous breeding throughout the colony. This is a phase of social facilitation that will be discussed at greater length in a later chapter.

These observations need to be extended and confirmed. They suggest one mechanism, that of mutual stimulation to mating, which may have operated to produce social nesting among birds, and which seems capable of giving added survival value to the larger colonies, once the habit of collecting into breeding flocks is established. We have here a suggestion that these social colonies of birds have evolved far enough so that there has come to be a threshold of numbers below which successful mating does not take place. The numbers that constitute this threshold vary under a variety of conditions.

In one case, when only two pairs were present, nests were built but no eggs were laid, whereas in a more favorable season, with three pairs, eggs were laid and one chick out of eight that hatched lived through the downy stage.

I saw the laughing gulls myself at Woods Hole. Years ago I found a paper by A. O. Gross giving the case of another almost extinct population which could not be revived. The heath hen, probably a representative of an eastern race of the prairie chicken, was formerly very abundant in Massachusetts, and may have been distributed from Maine to Delaware, or perhaps even farther south. It was gradually isolated by the killing of birds in the intermediate region and was driven back, until about

1850 it was found only on Martha's Vineyard and the near-by islands and among the pine barrens of New Jersey.[63] By 1880, except for attempted and unsuccessful introductions elsewhere, it was probably restricted to Martha's Vineyard. In 1890-92 it was estimated that one hundred to two hundred birds remained on that island. Then several things happened at about the same time: prairie chickens were introduced and probably interbred with the vanishing heath hen, protection of the birds was stiffened, and collectors' prices went up! It is an interesting commentary that most of the museum specimens, of which 208 are known at present, were collected between 1891 and 1900, when the probable extinction of the heath hen was noised abroad. This is one of the modern handicaps of small numbers; let a species or race become known to be rare, and museum collectors feel it their special duty to get a good supply laid in, just in case it does become extinct.

By 1907, when the Heath Hen Association was formed and employed a competent warden, the count had been reduced to seventy-seven. Massachusetts became aroused and purchased six hundred acres of heath hen range and leased a thousand acres more. The reservation was near a state forest that added another thousand acres of protected range. The birds responded to increased care and by 1916 it was estimated that there were two thousand in existence.

Then came a fire, a gale, and a hard winter, with an unprecedented flight of goshawks, and in April, 1917, there were fewer than fifty breeding pairs. The next year, when there was an estimated total population of 150, the heath-hen range was invaded by several expert photographers who took motion pictures of mating behavior. Even in the face of this disturbance at a critical time, a good year allowed the birds to increase and again to spread over Martha's Vineyard. In 1920, 314 were counted; but thereafter a decline in numbers set in which was never stopped. The figures for those five successive years are: 117, 100, 28, 54, 25. At this point extra wardens were put on the job, who killed more cats, crows, rats, hawks, and owls, the enemies of the heath hen. The next year's count was 35; in 1927, there were 20; but in 1928, in a census that lasted four days, only a single male was found. No other bird was seen thereafter, though a reward of a hundred dollars was offered for the discovery of another. This single male was banded and released and was last seen alive on February 9, 1932. With his death the heath hen became extinct.[22]

When this much is known of the decline in numbers of a given species there should be some knowledge of the factors involved in its extinction. There is. In the earlier years, as I have indicated with regard to museum collecting, there was undoubtedly a considerable amount of poaching; but as population of heath hens declined, local sentiment turned in favor of protection and poaching decreased, both because of a more intelligent public reaction to the birds and because of closer patrol by wardens. Dr. Gross, whose account I have been following, thinks that there was evidence of an inadaptability of the species, an excessive inbreeding, and, at the end, an excessive number of males. In such small populations the sex ratios frequently become highly abnormal. Disease and parasites took their toll. Predators, particularly cats and rats, were active. The females hid their nests well and were faithful in remaining on them, so that they were killed off by the fires that at times whipped over the breeding grounds.

Over sixty thousand dollars was spent in trying to save the heath hen, but without success. In contrast to the laughing gull, which nested in a relatively safe place and which came back from a population as low as the heath hen's until the very last, this unfortunate species was not able to adjust itself and continue existence, even with as intelligent human help as could be mustered in its favor.

The general conclusion seems to be that different species have different minimum populations below which the species cannot go with safety, and that in some instances this is considerably above the theoretical minimum of one pair.

By way of the laboratory, the coastal regions of Massachusetts, and South African grassland and forest, we are arriving at a general biological principle regarding the importance of numbers present on the growth, survival and, as we shall see, upon the evolution of species of animals.

Population
Size and Evolution.
Chapter VI

Charles Darwin recognized in 1859 in the *Origin of Species* that natural selection, to be effective at the individual level, needs a large population on which to work. Despite this early insight, precise information concerning the relations between population size and evolution is still scanty. Accordingly, we shall turn to mathematical explorations of its possibilities, as made primarily by Professor Sewall Wright.[46, 139] Although the ideas to be presented are essentially simple in principle, they are sufficiently novel and unfamiliar to challenge the closest attention.

I shall not indulge here in the details of the mathematical analyses, for the very good reason that I do not understand them. If I were not convinced, however, that Professor Wright does understand them I should not present this outline. It is only fair to say that, in my opinion, in dealing with these ideal populations Professor Wright cannot bring into sharp focus at one time all the factors that may be acting in nature. This is what he has been courageous enough to attempt; the more nearly he succeeds, the more likely is the calculation to be too complex for presentation in detail except to highly specialized readers.

The environment is in a state of constant flux, and its progressive changes, whether slow or fast, make the well-adapted types of the past generations into misfits under present conditions. The result may be rectified either by the extinction of the species, if it is not sufficiently plastic, or through reorganization of the hereditary types. In such a reorganization the simple Lamarckian reactions apparently do not operate; that is to say, when confronted with new, critical conditions, species cannot go to work and produce needed changes to fit new needs. The reactions are much more complicated than that.

To present the modern interpretation of this reorganization I need to use three technical terms which I shall first define. *Genes* are bits of protoplasm too small to be seen through the microscope, which are located in all cells and which are thought to be the bearers of heredity. They behave as indivisible units: that is to say, a gene if present in an organism is either transmitted as a whole or not at all. *Gene frequency* is the term applied to the frequency with which a given gene is found in a population, relative to the total possible frequency. By *mutation* is meant a large or small hereditary change that appears suddenly, usually in the sense in which I shall use it, as a result of a change in one or more genes. With these three terms in mind we are ready to try to understand how the hereditary types may become reorganized.

Such a reorganization implies a change in gene frequencies. By this I mean now that there will be a decrease in the abundance of the genes that were responsible for the past adaptations that are now obsolete, and an increase in the frequency of those genes allowing an adaptation to the new conditions. Gene frequencies remain constant in a large population unless changed by mutation, selection, or immigration. This is related to the unitary character, without blending, and the symmetry of the Mendelian mechanism of heredity.

These lifesaving genes may have been present in the species for a million years as a result of long past mutations, without having been of any value to the species in all that time. Now, under changed conditions, they may save it from extinction. It is important to note that organisms do not usually meet changed conditions by waiting for a new mutation; frequently all members of a species would be dead long before the right change would occur. This means that since a species cannot produce adaptive changes when and where needed, in order to persist successfully it must possess at all times a store of concealed potential variability.

I may interject parenthetically that at times this appears to call for the presence of a considerable number of individuals as a necessary condition to provide the needed variations. A part of this reserve of variability may be of no use under any circumstances; some characters may be useful; some may never meet with the circumstances under which they would have survival value; others, though of no use or even harmful when they appear, may later enable the species to live under newly changed conditions.

Hereditary changes tend to be eliminated as soon as they run counter to decided environmental selection. In large populations the results of mutations tend to stabilize about some average gene frequency, which represents the interaction between the rate of mutation and the degree of selection. Frequently, mutation pressure pushes in one direction and selection in another and the resulting gene frequency in the population represents a point or zone of equilibrium between these forces. In populations that are small but not too small, selection between genes becomes relatively ineffective, and the gene frequencies drift at random over a wide range about a certain mean position. In very small breeding populations, even though these may be small isolated colonies of a large widespread species, gene frequencies drift into

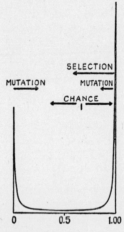

FIGURE 19. *In small populations, genes drift into fixation or loss largely irrespective of selection; the frequency of fixation or loss depends in the long run on the relative frequency of mutation and reverse mutation. (After Wright.)*

fixation of one alternative or another more rapidly than they are changed by selection or by mutation. Mutation, however, prevents permanent fixation. The condition at any given moment is largely a matter of chance.

Perhaps a diagram will help at this point. In Figure 19 the horizontal axis shows the different gene frequencies in a population, and the vertical axis gives the chances of the population under consideration possessing any given gene frequency. At the left, the gene frequency is zero; that is, the gene in question is absent from the population for the time being. The height of the curve shows that there is a good chance of this happening. At the extreme right the gene has become fixed and all animals in the population have it; they are a pure culture so far as this gene is concerned. Again there is a high degree of probability that this may happen when numbers are few. But the intermediate condition, when the gene is present in some but not all of the animals, shows little chance of occurrence.

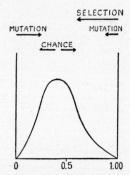

FIGURE 20. *In medium populations, gene frequencies drift at random about an intermediate point but not so much so that complete fixation or loss is likely to occur. (After Wright.)*

In such small populations, as has been said before, the gene frequency is determined mainly by chance; any given hereditary unit tends to disappear completely or become fixed and occur in all members of the small inbreeding colony. Such a condition may have been reached in the inbred population of the heath hen on Martha's Vineyard.

With populations that are intermediate in size there is a greater variety of possibilities. Some genes are lost, others reach chance fixations, and still others fluctuate widely in frequency from time to time. These conditions are shown in Figure 20.

If a given species is isolated into breeding colonies in such a way that but little emigration occurs between them—a condition known to exist in nature—in the course of time, as Professor Wright shows, the species will become divided into local races. This will happen although at the time of separation the populations were all homogeneous and the environment of all remains essentially similar.

If the environment does remain steady the larger colonies will tend to keep the same hereditary constitution as that which the whole species formerly had (Figure 21). Small breeding colonies will, however, become pure

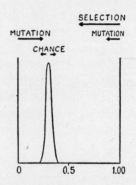

FIGURE 21. *In large populations, gene frequency is held to a certain equilibrium value as a result of the opposing pressures of mutation and selection. (After Wright.)*

cultures for different characters, and it is impossible to predict the course of the hereditary drift in any of these populations. As illustrated in Figure 19, the fixation will

be a matter of chance, and local races will result without
any necessary reference to adaptation.

The snails in the different mountain valleys of Hawaii
afford the classical illustration of this point. Each individual mountain valley has its separate species of snails. They
are distinguished by size, by color markings, and by other
characters that may be wholly nonadaptive.

Colonies that are intermediate in size will preserve a
part of the variability that will be lost in the smaller
colonies. Even so, there will be some independent drifting
apart of the various gene frequencies, so that these, too,
will give rise to new local races. Professor Wright's calculations show that with mutation rates of the order of
$1:10,000$ or $1:100,000$, such intermediate populations will
consist of some thousands or tens of thousands of individuals.

With small breeding populations, then, genes tend to
become fixed or lost. Even rather severe selection is without effect. Individual genes drift from one state of fixation to another regardless of selection. In large populations, gene frequencies tend to come to equilibrium between mutation and selection, and if selection is severe,
there tends to be a fixation of the gene or genes that
carry adaptive modifications, and evolution comes to a
standstill.

With a population intermediate in size, when there are
on one hand enough animals present to prevent fixation
of the genes, but on the other hand not enough animals to
prevent a random drifting about the mean values determined by selection and mutation, then evolution may
occur. The results obtained will depend upon the balance
between mutation rate, selection rate, and the size of the
effective breeding population.

In one more case the effect of differences in severity of

selection was worked out by Professor Wright (Figure 22). With a moderate mutation rate, if the selection is relatively weak, mutation pressure may determine the result and the given character will then drift to fixation or, as shown in the diagram, to extinction. As selec-

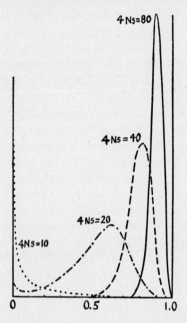

FIGURE 22. *As intensity of selection increases it becomes more and more dominant in determining the end result, and the degree of variation is lessened; 4Ns gives selection pressure. (From Wright.)*

tion pressures increase, selection tends to take charge of the end products, and, if slight, there is a wide variation about a mean; if more intense, the amount of variation becomes less and less.

When a species is broken up into different breeding

colonies, as it is with the snails in the Hawaiian valleys,[65] it can be similarly shown that the effects produced depend upon the rate of emigration between colonies, as well as upon selection pressure, mutation pressure, and population size, other factors being constant. Crossbreeding introduces genes into a population in a way that is essentially identical with mutation in its mathematical consequences; however, similar results may be obtained in a much shorter time by cross-breeding. In fact, all the different results which have just been illustrated can be duplicated by varying the numbers of the emigrants.

This is not the place to explore all the implications and possibilities of these interesting analyses. The highly significant conclusion has been reached that if a species occurs not as a single breeding unit but broken into effective breeding colonies which are almost isolated from one another, the members of different colonies, given sufficient vigor, may evolve into dissimilar local races. If one of these becomes well adapted to its environment it may increase in numbers and send out numerous emigrants. If these emigrants find and interbreed with members of other less advanced colonies, they will grade these up until they resemble the most adapted colony. This part of the process resembles a stock breeder's grading up of a mediocre herd of cattle by repeated infusions of new and improved "blood" into his herd. The significant thing here is that the random differentiation of local populations furnishes material for the action of selection on types as wholes, rather than on the mere average adaptive effects of individual genes.

The end results will vary even when the original population was homogeneous, and when mutation rates are similar throughout, even though selection is in the same direction in all parts of the different colonies. The primary factor under these conditions will be that of effective breeding population size, and there will be a somewhat greater chance for varied evolution among the populations that are intermediate in size, as contrasted with those which are small or large, and still greater chance for evolution when a large species is broken into small breeding colonies that are not completely isolated from one another.

This argument, even as I have simplified it, is not too easily followed the first time one goes over it. Perhaps my use of an old teaching trick, that of repetition of the same ideas with different words and different illustrations, may be forgiven. In doing so I am still leaning heavily on Professor Wright. The series of diagrams shown in Plate IV are built on one fundamental background. In perspective we see two elevations, one higher than the other, and two depressions that are the low points in a valley between the two peaks. Every position is intended to represent a different combination of gene frequencies. The peaks represent gene combinations that are highly adaptive; the depressions represent those that lack adaptive value. The degree of adaptiveness is shown by the height occupied by the given population. The variability of the population is shown by the size of the area that is occupied. Every individual in a species may have a different gene combination from every other, and yet the species may occupy a small region relative to all the possibilities.

We may call the lower peak Mount Minor Adaptation and the higher one Mount Major Adaptation. In A we find a population that is fairly well adapted, but not so much so as if it occupied the higher peak. Its original position and its variability are shown by the dotted circle. As a result of increased rate of mutation or of reduced selection, or both, the variability of the population has increased and it now spreads down to lower positions on

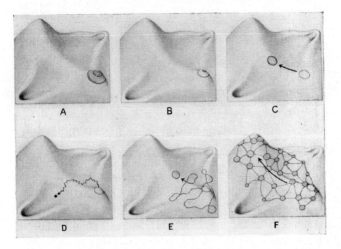

PLATE IV. *A population originally possessed a set of gene combinations of some slight adaptive value (dotted circle). With increased mutation rate it can expand to less adapted levels (A); with increased selection it contracts (B); if the environment changes the gene frequency must shift (C); with small numbers and close inbreeding the course of evolution is erratic and extinction usually follows (D); with larger numbers, evolution takes place more readily (E); most readily, when a large population is broken into local colonies with interemigration (F). (Modified from Wright.)*

this Mount Minor Adaptation. It contains more aberrant individuals and even freaks than when subject to less frequent mutation or to more severe selection, and a freak may appear that is more adaptive; but this important end has been achieved at the expense of the type of variability that might have made a major advance possible.

C introduces a different situation. As a result of environmental change, Mount Minor Adaptation has disappeared and the adapted population has been able to move to a new location at about the same level formerly occupied; now it is on the slope of Mount Major Adaptation, and if selection continues may be expected to move up that adaptive peak. A continually changing environment is undoubtedly an important factor in evolution.

The effects of population size are illustrated in the next three diagrams. The general background is the same as in A and B. In D is shown the effect of a decided reduction in population size, and consequently in variability, in the species that formerly occupied Mount Minor Adaptation. It is in fact so small that selection has become ineffective and the different hereditary qualities shift to chance fixations. As nonadaptive characters become fixed at random the species moves down from its peak over an erratic, unpredictable path. With reduction of population size below a certain minimum, control by selection between genes disappears to such an extent that the end can only be extinction.

With the species population intermediate in size, with the same mutation and selection rates as before, gene frequencies move about at random but without reaching the degree of fixation found in the preceding case. Since it will be easier to escape from low adaptive peaks, the population will tend finally to occupy the more adapted levels. The rate of progress is, however, extremely slow.

Finally, in F, we see the case of a species that has become

broken up into many small local races, perhaps as a result of restricted environmental niches. Each of these local races breeds largely within its own colony, but there is an occasional emigration from one colony to another. Each tends, if it is small in number, to give rise to different variations that shift about in a nonadaptive manner. The total number of relatively stable variations will be much greater since the total number of individuals is so much larger than in E. Under these conditions the chances are good that some of the local colonies will escape from the influence of Mount Minor Adaptation and manage to cross the valley to Mount Major Adaptation. Here the race will expand in numbers and will send out more and more emigrants which will interbreed with the stocks in the less adapted colonies and tend to grade them all up toward a higher adaptive level.

The conclusion is as Professor Wright says: "A subdivision of a large species into numerous small, partially isolated races gives the most effective setting for the operation of the trial and error mechanism in the field of evolution that results from gene combinations."

In the rate of evolution, therefore, population size is as important as we have seen it to be in the growth of individuals or in the growth of population numbers; and the optimal population size does not coincide with either the largest or smallest possible but lies at some intermediate point. Or, in Wright's words: "Conditions are more favorable in a population of intermediate size than in a very small or in a very large *random breeding* one (assuming a constant direction of selection)." Wright adds immediately that "conditions are enormously more favorable in a population which may be large but which is subdivided into many small local populations almost but not quite completely isolated from each other." [140] Evidently evolution is most favored by a large population much subdivided into almost isolated interbreeding elements as compared with either a small population or one so large as to prevent optimal subdivision into small, partially isolated breeding groups.

Group Behavior.
                Chapter **VII**

In the second chapter I told of how I stumbled on the fact that in the breeding season the normal behavior of isopods is affected by numbers present. Such effects have long been known for many types of behavior, and it would not be profitable here to catalogue and analyze all the cases that are on record. Rather, as before, I shall select certain well-authenticated examples of breeding reactions and of other types of behavior. Those chosen are especially noteworthy because of the behavior pattern that is involved, or because freshly observed, or both.

And here there will be a shift in emphasis. I have been stressing the existence of a widespread, fundamental automatic proto-cooperation that has survival value, and I have given evidence that it is a common trait in the animal kingdom. In this chapter I shall discuss group behavior that may or may not have immediate survival value. In each instance, and throughout the discussion as a whole, I shall be engaged in trying to find to what extent behavior is influenced by the presence of others, and I shall not consistently attempt to assay possible values that may or may not be involved.

With many more or less social animals the group up to a certain size facilitates various types of behavior. This is frequently called social facilitation. One phase of social facilitation is illustrated by some observations by the late student of birds Frank M. Chapman,[32] near the tropical laboratory on Barro Colorado Island, in the rain forest of Panama. Mr. Chapman found that males of Gould's manakin establish lines of courting places (Figure 23). The manakin is a small warbler-like bird, delicately colored and relatively inconspicuous. Each of the courting

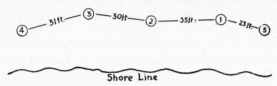

FIGURE 23. *Manakin males establish rows of mating courts in the Panamanian rain forest. (From Chapman.)*

places is occupied by a single male; the line thus formed extends for many yards through the undergrowth of the rain forest. From time to time each day during the long nesting season, the males resort to their individual cleared spots on the forest floor and make their presence known by a series of snaps, whirrs and calls that may be heard as far as three hundred yards away. The females, who are more quiet and retiring apparently are attracted by the line of males; they come individually from the surrounding thickets and each mates with one of the males. The evidence suggests that they are attracted from a greater distance by the spaced aggregation of males than they would be by isolated courting places. The more or less organized line of males in breeding condition apparently facilitates the mating of these jungle birds.

This is a highly specialized example of the widespread phenomenon of territoriality, which can be recognized even among breeding fishes[113] and which has been much studied of recent years in birds.[73] Typically, the male birds arrive first in the spring and take up fairly well-defined territories in the same general area, which they defend from intruding males. Then the females come in and flit from territory to territory before settling down to raise a brood with one particular male. There is always the strong suggestion that the presence of a number of singing males, even if spaced about in different territories, attracts and hastens the acceptance of some one of them by an unmated female.

Group stimulation of food consumption has been reported for various animals, including rats,[67] chickens,[27] and fishes.[130] I shall illustrate by some of the experiments conducted in our laboratory by Dr. J. C. Welty. These have been amply verified by other research workers. In connection with experiments on the effect of numbers on the rate of learning in fishes, which will be discussed later, Dr. Welty undertook to find whether grouped fish ate more or less than if they were isolated. The results of a typical experiment are illustrated in Figure 24.

Goldfish were photographed to scale, and those of similar size were selected for experimentation. Two groups of four each were placed in separate crystallizing dishes and each of eight others was isolated into a wholly similar dish. The different dishes were separated by black paper so that vision from one to the other was impossible. A known number of the small crustacean *Daphnia* were introduced daily into each dish. These living *Daphnia* had been screened so as to select the large animals only. As

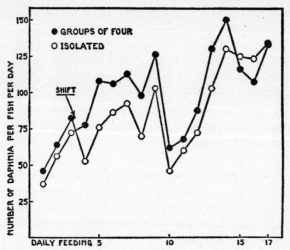

FIGURE 24. *Many kinds of fishes eat more if several are present than if they are isolated. (From Welty.)*

shown by the figure, fish in the groups of four ate decidedly more on the first three days of the experiment. At
this time the two lots were shifted. Those that had been
grouped were now isolated, and vice versa. There was an
immediate shift in the numbers of *Daphnia* taken, with
the newly isolated animals now eating less than the accompanying groups. This indicates that we are dealing
with an effect of numbers present rather than with
chance differences in individual appetites. This difference
kept up steadily until the last three days of observation,
when an interesting complication arose. By this time
each group of four fish was receiving a total of over six
hundred *Daphnia* daily, including those that were eaten
and the extras added to insure an economy of plenty. Each
isolated fish was in the presence of only one hundred fifty
Daphnia. Now six hundred and more large *Daphnia*, each
about an eighth of an inch long, make quite a swarm in a

none-too-large crystallizing dish. The consumption of food per animal by the grouped fish fell off. It was shown by appropriate tests that this resulted from the action of a so-called confusion effect. When fewer *Daphnia* were present, a fish might be observed to swim after an isolated crustacean and eat it, whereas a dozen *Daphnia* or so in the immediate field of vision seemed to offer conflicting stimuli that blocked the feeding response. Working on this suggestion, one group of four was given the usual quota of some six hundred *Daphnia* all at once; another group was given only one hundred at a time, and when these were approximately all eaten then another hundred would be introduced, and so on until the end of the regular feeding period. This prevented the *Daphnia* from being too dense at the beginning of the hour's feeding time. The isolated fish were fed as usual. Under these conditions the grouped goldfish that were fed one hundred *Daphnia* at a time ate definitely more than those given the whole confusing mass at once.

Here we come upon not one but two mass effects. In the first place we see that the fish in groups of four were stimulated to eat more food than if isolated, and that this depended on their state of aggregation. But, incidental to this demonstration, we find that in the presence of too many animated food particles a confusion effect arises that decreases the feeding efficiency of the fish.

It has been suspected for years that such a confusion effect exists and has survival value for small animals flocking together in the presence of a predator, such as small birds in the region of a hawk. These observations of Welty's make the best demonstration that I know of the existence of such an effect, in this case the *Daphnia* in the presence of the fish. I am less interested in this confusion effect at present than in the demonstration of social facilitation in feeding, a phenomenon that has been shown

to exist for a number of fishes, including zebra fish, para-
dise fish, goldfish and guppies of the more usual aquarium
varieties, and the lake shiner, *Notropis atherinoides*, as
well.

None of these fishes is very social, that is, none of them
groups into close schools. For evidence of similar social
stimulation among social animals it is interesting to examine the effect of numbers present on the digging behavior of the highly social ants. The account of this work
was published in 1937 by Professor S. C. Chen, of Peiping,
China.[33]

These ants, a species of *Camponotus*, dig their nests
in the ground. It was found that all the worker ants of
this species are capable of digging a nest when in isolation,
but that the rate of work varies with different individuals. If marked ants, whose reaction time has been tested
in isolation, are placed together in pairs or in groups, they
will start work sooner and will work with greater uniformity than if alone.

Professor Chen and his assistants collected and counted
the number of the tiny pellets of earth that were dug by
different individual ants when isolated, and when members
of groups of two or three ants. They found that the
number of pellets removed is greater when the ants work
in association with others than when each works alone.
This accelerating effect is greater for slow than for rapid
workers; when ants with intermediate working tendencies were tested, (Figure 25) they were found to be
speeded up when in company with a rapid co-worker and
relatively retarded when placed with a slow-working ant.
Interestingly enough, there was no difference between
the stimulating effect of one additional ant and of many
ants on the rate of work of a given individual. The social
facilitation seemed maximal for these digging tests when
only a second individual was present.

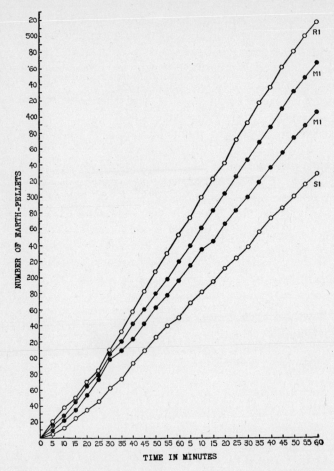

FIGURE 25. *An ant that works at an intermediate rate (M1) may be speeded up if placed with an ant that works more rapidly (R1) and slowed down if put with a slower worker (S1). (From Chen.)*

Ants that regularly work rapidly were found to be physiologically different from those that work more slowly. The faster workers were more susceptible to starvation, to drying, and to exposure to ether or to chloroform. Tests that have been made by others indicate that animals that are more active physiologically usually succumb sooner under such adverse conditions, just as these rapidly-working ants were found to do. These are exceedingly interesting results because here we see that ants with apparently innate differences in speed of fundamental processes are affected in their speed of digging by the presence or the absence of a nest mate. The ant of intermediate speed, presumably with an intermediate underlying reaction system, is most interesting of all, because it can be either speeded up or retarded according as it is placed with an active or a more passive individual.

In this connection we have known for over two decades scientifically what was common sense before that time, namely, that human animals, whether adults or children, can accomplish more mental and physical work, at least of certain kinds, and will work with greater uniformity when in association with others doing similar tasks, than if obliged to work in isolation.[19, 95]

Such considerations lead directly to problems concerning the effect of numbers present on the rate of learning in man. Here we find a set of questions with great and immediate human significance. The world over, the training of the young of their own species is one of the major preoccupations of mankind. This is particularly true in the United States, where we are engaged in mass education on an unprecedented scale. This teaching of the young to the extent to which we are attempting it is an expensive business in time, in effort, and in money. We need to know, therefore, the number of these interesting young animals that can be trained together with best results. In other words, what is the optimal class size for the various levels of training from preschool days through the preparation for the doctor's degree and further?

In part, the proper answer to this question calls for a statement of educational objectives. The development of strong individuality, for example, is not necessarily accomplished by the same teaching methods and class size that favor the growth of conformity to group patterns; and the rapid development of mastery of so-called skills may call for different number relations than those needed for the mastery of logical thought.

Even without positive information we can guess that the tutorial method with individuals or very small groups will best serve certain ends, whereas others will be achieved most readily in larger groups. The question, or a simplified part of it, thus becomes: What class size favors optimal rate of learning of the usual class material presented at different ages?

As might be anticipated, the difficulties of human experimentation being what they are, it is hard to collect accurate information on this point. Much depends upon the comparative accuracy of the sampling, and also on more subjective factors, such as the attitude of the teacher and of the students toward large and small classes. There is also a factor that I have not seen mentioned in the literature on the subject, the effect on the student of his realizing or suspecting that he is an object of experimental interest, an educational guinea pig. This stimulus is more likely to be potent, in my opinion, when the student is a member of a class that is unusual in size.

In the more careful studies, results of which have been published, the class numbers have ranged from "small" through "medium" to "large." The "small'" experimental classes apparently have about twenty to twenty-five members; this represents a more usual experience to the student, and he is more likely to be conscious of class size when he is a member of a large class of seventy-five or more than when he is in a small class or a medium-sized one of thirty-five to forty. The sizes that are counted "large" or "small" vary greatly, sometimes in the same experimental treatment, so that frequently the comparisons are between larger and smaller classes, both medium in size, rather than between real extremes in numbers.

Frequently, too, the teaching practice varies in the two classes. Thus in one experiment the smaller classes in high-school geometry contained about twenty-five members and the large ones had about one hundred. In the large classes a student helper was present for every ten class members. These helpers were superior students in geometry of the preceding year. As nearly as I can discover, there were no student helpers in the small classes. Under the conditions it is perhaps not unexpected that a better showing was made by those in the large classes. With them, there were present not only more instructors per student but these were people of nearly their own age, who could be approached without hesitation not only in class but out of class and even out of school hours. Every mature teacher knows that even with the best intention and the most democratic attitude, age differences widen the gap between the teacher and the taught, whatever other compensations there may be.

The most comprehensive experiments I have seen reported in this field are those of the subcommittee on class size of the committee on educational research at the University of Minnesota.[74] These were carried on at the college level and involved one hundred nine classes under twenty-one instructors in eleven departments of four colleges in the University of Minnesota. In large classes, 4,205 students were observed, and in small ones, 1,854; of these, 1,288 were paired as to intelligence, sex, and scholarship before the experiment was begun. One of each pair was assigned to a large and one to a small class in the same subject taught by the same instructor. In this way the obvious variables were controlled as well as is humanly possible, unless we could have a large number of identical twins with which to experiment.

In 78 per cent of the experiments a more or less decided advantage accrued to the members of the pairs who were in the large classes, and at every scholarship level tested the members of the pairs who were in the large sections did better work than the members of the pairs in the smaller ones; the excellent students appeared to profit somewhat more from being in large classes than their less outstanding fellows.

Of the available data, a re-examination of the summaries indicates that there is on the average a difference in the means in the final grade of 4.1 points, favoring the students in the larger classes. There is a statistical probability of matching this by random sampling of four chances in ten million (P=0.0000004), and this despite the fact that the majority of the class comparisons did not give significant differences until all the evidence was considered together.

The numbers in the smaller classes usually ranged from twenty-one to thirty, but in some classes dropped as low as twelve; in the larger classes there were usually thirty-five to seventy-nine students; in the largest, one hundred and sixty-nine. Under the conditions that prevailed in these classes in psychology, educational psychology, and physics, the students in the larger class sections made slightly but significantly higher final grades than those in smaller sections of the same subject taught by the same instructor.

So much for objective experiments. It happens that subjective estimates, made both by teachers and by students at Minnesota, favor the smaller rather than the larger classes. It was even true that the students were better satisfied with the marks received in smaller classes than they were with the slightly higher grades given them in the larger sections.

The general attitude seemed somewhat like that toward a friend of mine who teaches general mathematics at Purdue University. He is an experienced and excellent teacher. His program for one semester required that he should meet a normal-sized class of thirty to thirty-five at eight o'clock, and that at nine o'clock he should meet a class of double the size in a larger room, to repeat the same subject matter. At the close of the semester the two sections were asked to rank the instructor on many different points. Uniformly the students in the larger section rated him lower than those in the smaller section, in such matters as teaching skill, pleasantness of voice, neatness of appearance, and personal attractiveness!

I have had a fairly extensive teaching experience, which has included work in grade- and high-school teaching, as well as over thirty-five years of teaching at the college and university level, during which time I have taught classes of almost all sizes, from those of over six hundred at the University of California to the graduate classes of three or four that come my way; and I must confess to a personal prejudice against these very large classes. Even when using the same lecture notes, I do not give the same lecture to five hundred students that I give to forty or fifty. On the other hand, even with graduate classes and advanced seminars I am prejudiced in favor of having enough students, which means at least eight, to give a certain *esprit de corps* to the group. Such personal opinions have their value, particularly when they click with experimental results such as those outlined by Hudelson from the experiments at Minnesota. It is unfortunate that those experiments did not test either the upper or the lower limits of class size which are conducive to good classroom performance on the part of the students; I know of none that does test these points adequately.

Some of the difficulties inherent in experimentation on the effects of class size on the rate of learning in man can be obviated by the use of nonhuman animals. This procedure does not solve all the requirements for elegant objective experimentation, and it has the additional real difficulty of eliminating all possibility of adding subjective impressions to objective findings, a point that makes one of the strongest arguments for experimentation on man when feasible.

In some respects the most completely controlled experiments on the effect of numbers present on the rate of learning are those that Miss Mary Gates and I performed some years ago, using common cockroaches as experimental animals.[58] Earlier work by two independent investigators had shown that cockroaches can be trained to run a simple maze and can show improvement from day to day. In our experiments we found that the cockroaches could be trained to run the maze we used by fifteen to twenty-five successive trials on a given day, and that they showed definite improvement both in time taken to run the maze and in number of errors. However, unlike the experience of our predecessors, these University of Chicago cockroaches could not carry over the effects of training from one day to the next.

The reason for this difference between our cockroaches

and those around St. Louis and in Germany is not known. It may be that at the University of Chicago, despite our reputation for scholarship, the local cockroaches have a low I.Q., or it may be that since we used animals from the bacteriological laboratory, because of their unusual size and physical vigor, we were unconsciously selecting the dumber sort. Or perhaps, contrary to our plan, we set them a problem that is intrinsically more difficult for the cockroach mentality. In any event, it is important to remember that our cockroaches forgot overnight anything they may have learned the day before. As it turns out, this was fortunate for the experiments we were carrying on, because we could match up individual cockroaches with the same speed of learning in pairs or groups of three for later tests without fear of a carry-over from their previous experience.

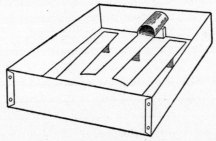

FIGURE 26. *A simple maze used in training cockroaches.*

The maze used is shown in Figure 26. It consisted of a metal platform from which three runways extended, each about two inches wide and a foot or so long. The two side runways ended blindly, but the center one led to a black bottle, which allowed the cockroaches to escape from the light. This apparently was a reward for cockroaches which, when possible, give a negative reaction to light.

The three-pronged set of runways was mounted about half an inch above a pan of water; the majority of the cockroaches tended to avoid it and so kept on the runways. The tests were all made in a dark room and light was furnished by a single electric bulb mounted just above the point where the central runway left the main platform. In other words, the cockroaches, which are negative to light, had to learn to run through the area of strongest illumination in order to reach the dark bottle that served as a reward. After two minutes' rest in the dark bottle the cockroaches were literally poured out onto the platform of the maze without being touched by the experimenter, and observation of them began again.

The problem as set was about at the limit of cockroach ability. Approximately one third of the insects tested never learned to stay on the maze; whenever they were placed on it they proceeded immediately to run off into the underlying water. Of the two thirds that did learn to remain on the maze, a half, or another third of all those tested, did not show improvement in speed of reaching the bottle after repeated trials. Thus only one third of the cockroaches we tested showed improvement with experience, and, as I said before, they forgot overnight all that they learned during the day.

As shown in the summarizing graphs (Figures 27 and 28), isolated cockroaches made fewer errors per trial throughout the whole training period. They also took less time to run the maze than when the same animals were members of pairs or of groups of three. Turning the comparison around, paired cockroaches took longer time per trial and made more errors than when isolated, and groups of three took still longer and made more errors than those in pairs.

A study of the rate of improvement shows that during

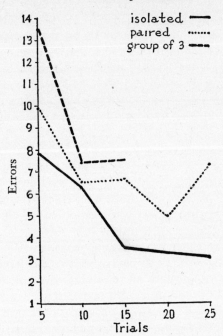

FIGURE 27. *Isolated cockroaches make fewer errors on the maze than the same animals paired, and still fewer than if three are being trained together.*

the early part of the training particularly so far as errors made are concerned, paired cockroaches improved more rapidly than they did if isolated or in groups of three, and those placed three together on the maze improved somewhat more rapidly than they did when isolated. Thus, although the presence of one or two extra cockroaches slowed down the speed of reaction on the maze and increased the number of errors made at all times, the rate of improvement was higher when more than one was present.

Excluding this one aspect of rate of improvement, in

112

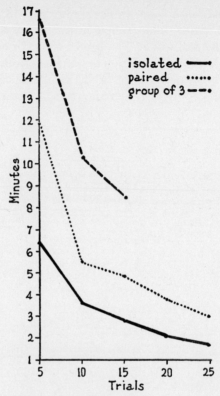

FIGURE 28. *They also take less time.*

all other phases of the experiment isolated cockroaches turned in better learning performances than they showed when more were present. Evidently under the conditions of our experiments the tutorial system usually works best with cockroaches.

Essentially the same sort of experiment was tried with isolated and paired Australian parakeets, which are commonly called lovebirds.[15] Rather naïvely, perhaps, I thought that, since these birds so readily pair off, perhaps two might learn to run a simple maze more rapidly than a single individual would. This turned out to be entirely a mistaken idea. I shall spare you the details concerning this maze; it was adequate in size, so that two birds could pass through practically abreast. Almost all the ninety-odd

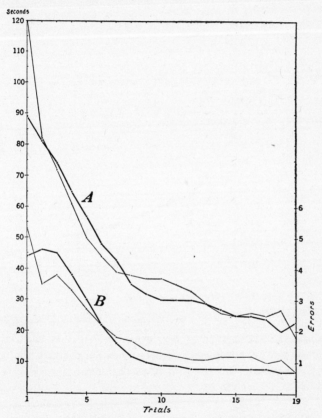

FIGURE 29. *Parakeets learn equally well if trained when isolated, whether they are caged singly or in pairs. A, time per trial; B, errors per trial.*

birds that were tested learned easily to run the maze and normally reduced their time per trial from about two minutes to a few seconds, after six or seven days of training. Errors also were reduced, and several of the birds were trained so that they ran the maze day after day with no errors at all.

The summarizing graphs Figures 29 and 30 outline the results obtained. It made no difference whether the birds were caged in pairs or separately; if placed alone in the maze the performance was similar. If, however, two birds were put together in the maze, the speed was reduced and errors increased as compared with the scores made by

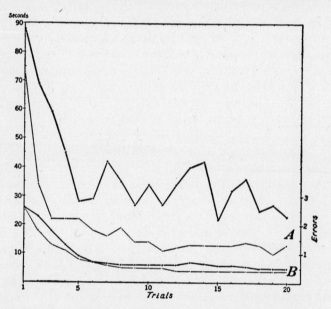

FIGURE 30. *Parakeets learn more rapidly if trained alone than if two are placed together in the maze. A, time per trial; B, errors per trial. (The uppermost curve is unsmoothed; the lower three have been smoothed mathematically.)*

isolated parakeets. When two males, two females, or a male and a female were trained in the maze together, there was always interference. The tendency was for the more rapid bird to slow down rather than for the slower bird to speed up. Both members of a pair of birds tended to take the same time and to make the same errors. Given sufficient training, they might make perfect scores so far as errors were concerned, but even after long training the performance of pairs was always more erratic than that of isolated birds. When birds that had been trained to a consistent level of excellence were exchanged so that those formerly isolated were paired and those formerly paired were isolated, their behavior in the maze took on the characteristics usually shown by paired and by isolated birds, proving that the type of reaction given was a result of the numbers present rather than of the working of other factors. With these lovebirds then, contrary to the original assumption, all indications were that being paired in the maze slowed down the rate of learning and increased the erratic character of their behavior.

Our experience with the general problem did not end here. I taught at the University of Chicago a favorite course called Animal Behavior. In this class the beginning research students attempt some small problem and frequently make good progress toward its superficial solution. One of these student projects has been the training of the common mud minnow to react to traffic lights. The fish were trained to jump out of water and obtain a bit of earthworm when red was flashed. Under the green light they were conditioned to retire to one of the bottom corners. If they did jump under green light they were fed filter paper soaked in turpentine. Within two months a lot of fishes, isolated one in each small aquarium, could be trained so that they would have been given an

A for the project if they had been properly enrolled students.

When, however, several fishes were placed together in the same aquarium and an attempt was made to train all at the same time, the rate of learning was retarded. Paired fish reacted as well as if they had been isolated, but the reactions of groups of four were slowed down, and those of ten even more so. Two fish would rarely jump at once, and when some one individual was getting set to jump for the food under the red light, another would frequently come along and give him a jab in the belly that would stop all tendency to jump for the time.

One more instance remains to be reported. Dr. Welty, who has been mentioned before, undertook to train goldfish to move forward from the rear screened-off portion of an aquarium through a door into a small forward

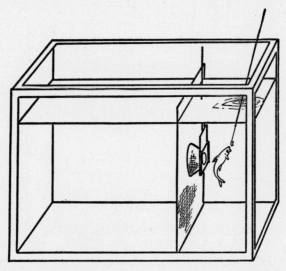

FIGURE 31. *Feeding a fish that has just come through the opening from the larger side of the aquarium. (From Welty.)*

chamber where each was fed just after it came through the opening.[130] A divided aquarium, similar to those used, is shown in Figure 31. The signal to the fish that it was time to react came from increasing the intensity of light in the aquarium and opening the door between the two compartments. Under Dr. Welty's careful coaching the fish improved rapidly in their speed of reaction and usually had reached a good level of performance by the sixth day of training.

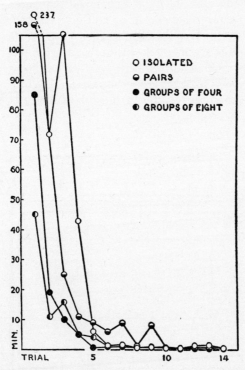

FIGURE 32. *Goldfish learn to perform in a simple divided aquarium the more readily the more fish there are present. (From Welty.)*

In his experiments almost a thousand fishes were trained at one time or another. The results of a sample experiment are recorded in Figure 32. In this test there were eight goldfish, each isolated in individual aquaria, four sets of paired goldfish, two lots of four placed together, and one group of eight in one aquarium. As shown by the graph, there was a marked group effect on the rate of learning. The speed of first performance of the untrained fishes was most rapid with eight present and slowest with isolated goldfish. In the early days of rapid learning the same order held. This experiment was repeated several times with identical results. Under these conditions there seems to be little doubt that the groups of goldfish learned to move forward and secure food more rapidly than the same number of isolated fish.

The conditions of the experiments allow certain types of analyses to be made. One of these is to test the te-

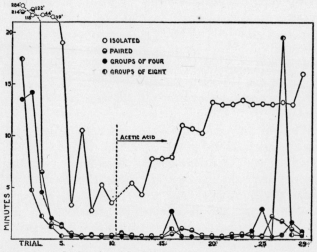

FIGURE 33. *Isolated goldfish learn the problem set for them less rapidly, and unlearn it more readily. (From Welty.)*

nacity with which the newly acquired habit will be retained. A set of fish was trained as usual (Figure 33). After ten days, when the grouped fish had been letter-perfect for four days, although the isolated goldfish were still taking some three minutes per trial, the experiment was changed; whenever the fish came forward through the gate they were offered pieces of worm soaked in acetic acid. The isolated fish, perhaps because they had not learned to perform so well, perhaps because they were isolated, or for some other reason, ceased to react rapidly, and on the twenty-ninth day they were averaging fifteen minutes per trial. The grouped fish were much more steady in behavior, and persisted in coming forward with relatively little change until the twenty-seventh day; and even then the old conditioning held for most of the fish most of the time. Many individuals persisted in coming forward through the gate for a long time after they ceased biting or even swimming toward the acid-treated worm.

When a group of fish are reacting together, if a given individual moves forward through the gate to the feeding space, others may follow because of a group cohesion. It is obvious that if a fish is isolated and moves forward the faster reaction cannot affect the behavior of other isolated fish.

With this in mind, Dr. Welty undertook a series of experiments in which there were two partitions in the aquarium, with one door opening forward and another door opening through the other partition toward the rear of the aquarium (Figure 34). The fish were placed in the central space and those in half the tanks were trained to come forward as usual. In the other half, two selected fish were conditioned to come forward and two were similarly trained to move to the rear compartment to be fed. The experiment was tried several times with goldfish,

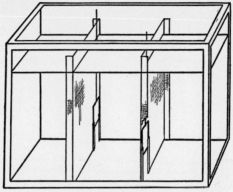

FIGURE 34. *The divided aquarium used in training part of the fish to come forward and part to go to the rear to be fed. (From Welty.)*

the minnow *Fundulus*, common at Woods Hole, and another marine minnow, *Cyprinodon*. For one reason or another, only one series in which the fish were closely comparable with each other was successfully completed. The results are shown in Figure 35. Generally speaking, the cohering groups of *Cyprinodon* learned more rapidly and reacted more steadily than the separating groups. Group cohesion, then, is one factor that is working, at least at times, in causing grouped fish to learn more rapidly in a simple divided aquarium than isolated fishes under similar treatment.

As the goldfish move forward in the usual divided aquarium there comes a time when one or more fish may be in front of the screen and the others in the rear of this advance guard. It was obviously a part of the investigation to find the effect these more rapidly reacting fish had upon their fellows merely as a result of being in the forward chamber. Conceivably they may have served as a lure. Another possibility is that a rapidly learning in-

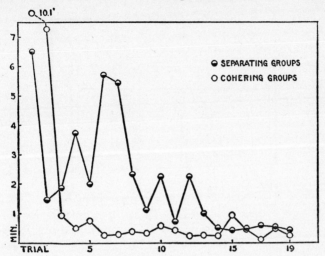

FIGURE 35. Cyprinodon *learn to move in a body more read-ily than to split into two separate groups. (From Welty.)*

dividual becomes a leader in the reaction of the whole group.

Both of these possibilities were tested experimentally by Dr. Welty, with results that are summarized in Figure 36. Three sets of aquaria were established. In the control aquaria all the goldfish, of which there were four in each tank, had had no previous experience in these experiments. These fish were trained as usual. In another set, an un-trained fish was kept in each forward compartment as a lure and four untrained fish were placed in the rear com-partment. These fish were trained as usual; the so-called lure-fish was fed after the first of the untrained lot came through the gateway. In the final set of aquaria a trained fish was introduced along with the four untrained fish. When the light was admitted and the gate was raised this trained fish moved forward, came through the gateway, and was fed immediately. The others followed.

As the graphs show, after the first day there was little difference between the reactions of the control fish and those of the fish that had a lure-fish in front of the screen. The fish with a trained leader generally gave more rapid reactions than either of the others.

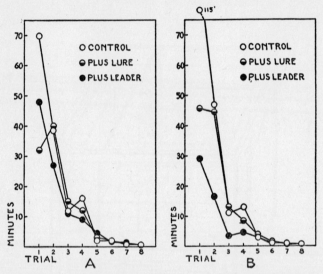

FIGURE 36. *Goldfish learn more readily if accompanied by a trained leader than if there is a fish in the proper position to act as lure. (From Welty.)*

There is always a temptation to make comparisons between the learning behavior of these laboratory animals and that of men. Direct comparisons should usually be avoided. However, in human terms, the goldfish reacted more rapidly in the presence of a trained leader that went through the whole behavior process with them than they did to the presence of one of their kind as a lure-fish in the forward compartment, a sort of signpost to proper behavior. Evidently leaders swimming with these goldfish can influence them more than fish which by their position merely show them where they can come. It seems fair to say that, with these fish, demonstration teaching is the most effective method yet discovered.

Still another attempt was made to study group cohesion in these goldfish. For this purpose aquaria were

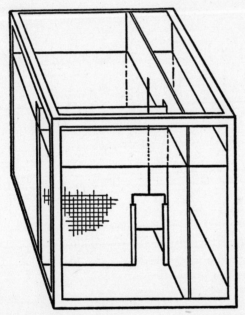

FIGURE 37. *A divided aquarium arranged to test the power of observation of fish placed in the side compartment. (From Welty.)*

arranged like those in Figure 37. At the side of the usual divided aquarium a narrow runway was placed into which untrained goldfish were introduced. In half of the tanks the glass partition was clear and allowed these fish to see the reaction of those in the larger chamber. In the other half of the tanks the partition was of opaque glass, cutting off the view.

Trained fish were placed in the larger chamber and were run through their performance from ten to twenty times in different experiments. The same treatment was given the fish in the aquaria with opaque partitions and those

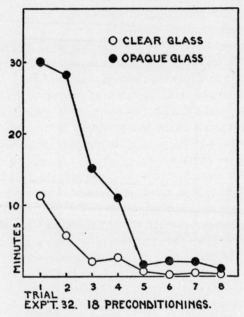

FIGURE 38. *Goldfish react more rapidly if allowed to watch others perform. (From Welty.)*

with clear glass. The trained fish were then removed and those from the small side chamber were gently transferred to the larger side. An hour later they were given an ordinary test such as had been given to the trained fish. As is clearly shown by the graphs in Figure 38, the fish that had been able to watch the others react behaved decidedly more like trained fish than those that had not been able to see their fellows perform.

As a final check, the whole test was repeated, except that no fish were placed in the larger side of the aquarium. Fifteen times each aquarium was lighted up, the door opened, and the experimenter stood ready to feed any imaginary fish that might come through. When those in the side passages were transferred, there was no essential difference in the behavior of the fish from that of untrained fish, and the experimenter was free from any suggestion that he might have been signaling the fish.

The results of these experiments suggest that there is such a thing as imitation among goldfish. Whether there is or not depends, as Dr. Welty rightly says, largely upon the definition given to the word imitation. These fish probably do imitate each other on a relatively simple level. The untrained fish that watched the reaction of their trained fellows through the clear glass became conditioned in two ways which were not open to the fish behind an opaque glass. In the first place, they saw the fish move forward on the reception of a given stimulus, pass through the gate, receive food, and give no evidence of an avoiding or "fright" reaction. This probably gave what might be called a certain reassurance. Secondly, they showed group cohesion, and moved forward as the reacting fishes did so; at times they were even seen to move forward in advance of the fishes on the larger side of the aquarium.

When transferred to the simple divided aquarium and given the releasing stimulus of an increase in light, accompanied by the opening of the gate, both types of previous experience probably played a role in producing a faster reaction. Fish behind the opaque glass could have neither of these helpful experiences. When their narrow aquarium was flooded with light they ordinarily moved back to the far end and remained there. There was nothing to train them to overcome this normally negative reaction. So reviewed, it must be said that this behavior has some points of resemblance to what is called imitation in other animals.

There is also an element of imitation in the greater food consumption of grouped fishes. One fish sees another pursue, attack, and consume a bit of food and its own feeding mechanism is set off as a result of this visual experience, even though its own hunger might not have been sufficient to stimulate feeding behavior. It is difficult to say to what extent such behavior is an expression of competition as contrasted with unconscious cooperation. The two types of motivation overlap here and elsewhere.

The evidence that we have been considering furthers our understanding of the fundamental nature of group activities among many animals, some of which are not usually regarded as being truly social. The whole emphasis of this chapter has been laid upon facilitation as the result of greater numbers being present. This kind of social facilitation has been described for such diverse processes as breeding behavior, eating, working, and learning.

Added numbers do not always facilitate these activities, as was shown by the analyses of the effect of numbers upon the rate of learning. With some animals—for example, men and goldfish—under certain situations learning is more rapid with several present; but with other animals, such as parakeets and mudminnows, under the conditions tested, increased numbers lead to a lower rate of learning. It seems that no all-inclusive positive statement can as yet be made in this field. One can, however, make the affirmation that in the general realm here considered the presence of additional numbers by no means always retards, and is frequently stimulating. As we saw before with regard to other processes, in certain cases there are ill effects of undercrowding as well as ill effects of overcrowding. Without careful experimental exploration, we cannot predict which effect will emerge from a given situation.

One other result comes from these studies that will help us to clarify evidence still to be presented, as well as to review that already given. We have come upon another measure of the existence of social behavior. Reactions may be regarded as social in nature to the extent that they differ from those that would be given if the animals were alone. Such differences are frequently quantitative, as they have been in the cases we have discussed, although qualitative differences occur as a result of a change in the numbers present.

From this point of view social behavior may have or may lack positive survival value. To be considered social, all that is necessary is that the behavior be different from that which would be shown if the animal were solitary. In this sense all the animals whose behavior we have been discussing are social to a considerable degree; the more so the greater the difference between their behavior when grouped and when isolated.

When the behavior of such animals as cockroaches, fishes, birds, and rats shows evidence of distinct modification as a result of more than one being present, we have another suggestion that there exists a broad substratum of partly social behavior. There are many indications that this extends through the whole animal kingdom. From such a substratum, given suitable conditions, societies emerge now and again as they have among ants and men. At these higher social levels, as is to be expected, the type of behavior shown under many conditions is re-lated even more closely to the number of animals present than with less social cockroaches and fish.

Group
Organization.
Chapter **VIII**

We all know that human society is
more or less closely organized. Sometimes, as in military
circles, some business organizations, and certain universities, there is a line organization which extends in
a definite order, step by step from the highest official to
the lowest rank. Frequently, however, the organization
is more complex, intricate, and temporary.

We have known for some time, too, that in herds of
the larger mammals, where one can distinguish different
individuals, the group may be organized to some extent
with a dominant leader and frequently with subleaders
that stand out above the common run of the herd.[20]

Despite this knowledge, we have found with surprise
that other animal groups, such as a flock of birds, in which
the different birds are indistinguishable to the human
eye, also are organized into a social hierarchy, frequently
with a well-recognized social order that runs through the
entire flock. The situation in these flocks of birds is
amusing, interesting, and important enough to warrant
more attention than it is receiving at present.

Studies of the sort I am going to describe were initiated by a Norwegian named Schjelderup-Ebbe.[118] They were made possible by the use of colored leg bands and other markings by which the different individuals could be recognized by a human observer. Apparently the birds themselves knew the individual members of the flock without such artificial aids.

Not because they represent the most important work on the subject, but because I can best vouch for them in detail and in general, I shall present certain analyses of group organizations that have been made in our own laboratory.

The organization of flocks of chickens is fairly firmly fixed. This is particularly the case with hens. The social order is indicated by the giving and receiving of pecks, or by reaction to threats of pecking; and hence the social hierarchy among birds is frequently referred to as the peck order.

When two chickens meet for the first time there is either a fight or one gives way without fighting. If one of the two is immature and the other is fully developed, the older bird usually dominates. Thereafter when these two meet the one that has acquired the peck right, that is, the right to peck another without being pecked in return, exercises it except in the event of a successful revolt that with chickens, rarely occurs.

The intensity of pair contact-reactions varies greatly. A superior may peck a subordinate severely, or lightly, or it may only threaten to do so. It usually turns its head, points its bill toward the subordinate, and takes a few steps in that direction. It may then give a low deep characteristic sound that frequently accompanies an actual peck, and stretch its neck up and out without the resulting peck that it seems just ready to administer.

The peck, when actually delivered, may be light,

heavy, or slashing. These vigorous pecks may be painful even to man, as anyone can testify who has tried to take a setting hen off her nest, and particularly painful if repeated in the same spot. The pecking bird may draw blood from the comb or may pull feathers from the neck of the pecked fowl. The peck is frequently aimed at the comb or the top of the head; often it is not received with full force, for the pecked bird dodges. Less often the peck is directed toward back or shoulders.

The severity of a peck that lands as aimed is illustrated by a recent observation in one of our small flocks. One bird received a vicious peck directly on the top of its head; it walked backward two or three feet, staggered and fell, arose and again walked backward in a blind course that took it into the bird that had given the original peck. By that time the aggressor had turned to eating and paid no attention to this chance contact.

RW pecks all 12: A, BG, BB, M, Y, YY, BG_2, GR, R, GY, RY, RR.
RR pecks 11 : A, BG, BB, M, Y, YY, BG_2, GR, R, GY, RY.
RY pecks 10 : A, BG, BB, M, Y, YY, BG_2, GR, R, GY.
GY pecks 9   : A, BG, BB, M, Y, YY, BG_2, GR, R.
R   pecks 8   : A, BG, BB, M, Y, YY, BG_2, GR.
GR pecks 7   : A, BG, BB, M, Y, YY, BG_2.
BG_2 pecks 6 : A, BG, BB, M, Y, YY.

YY pecks 4 : A, BG, BB, M.
M   pecks 4 : A, BG, BB, Y.
Y   pecks 4 : A, BG, BB,   YY.

BB pecks 2 : A, BG.
BG pecks 1 : A.
A   pecks 0

FIGURE 39. *Flocks of hens are organized into a definite social hierarchy.*

As a result of patient watching of pecks received and delivered, it is possible to find, with a high degree of accuracy, the social status of birds in a relatively small flock.[90] The organization of one such flock of brown leghorn pullets is shown in Figure 39. This peck order was determined after sixty days of observation. As shown by the chart, there was a regular line organization down to the eighth bird. Then a triangle was encountered in which M pecked Y, Y pecked YY and YY pecked M; and each of these had the peck right over the remaining members of the flock.

'   Such irregularities are by no means uncommon even in well-established flocks. A hen that is otherwise the alpha bird in the pen may be pecked with impunity by some low-ranking member, although the latter is in turn pecked by many birds over which the alpha hen has a clearly established social superiority. This inconsistency may result from the low-ranking bird having first met the alpha bird on one of its off days, gained the advantage in the first combat and managed to keep it thereafter with the aid of the psychological dominance thus established.

Similar social hierarchies exist also among flocks of male birds. One flock of cockerels, which we studied for seventy days, demonstrated the social order shown in Figure 40, in which there are six triangle situations that run through all the upper part of the social scale, but are especially evident in the middle ranks, where B is involved in four of them.

Cockerels are more pugnacious than pullets, even when they are kept as these were, on a diet that somewhat restricts the tendency to fight. There were more revolts than pullets make against those above them in the peck order, and these were more likely to be successful. For example, in this flock of cockerels, the four birds identified in bold-faced type in Figure 40 showed reversals, and with some the social rank had not been finally determined even after seventy days of observations. Thus BY was ob-

BW   pecks 9: W, BY, G, RY, B, BG, Y, R, GY.
BR   pecks 8: W, BY, G, RY,   BG, Y, **R,**   BW.
GY   pecks 8: **W**, BY, G, RY, B, **BG,** Y,   BR.
R   pecks 7: W, *BY,* **G,** RY, B, BG,   GY.
Y   pecks 6: W, BY, G, RY,   BG,   R.
GB   pecks 5: W, BY, G, RY, B.
B   pecks 4: W,   G, RY,   Y.
RY   pecks 3: W, BY, G.
G   pecks 2: W, *BY.*
BY   pecks 2: W,   B.
W   pecks 0.

In this order there are six triangle situations as follows:

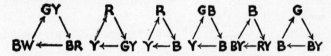

FIGURE 40. *Cockerels also have a social organization; it is, however, somewhat more confused than that of hens.*

served to peck G on six occasions, and G to peck BY eight times. Ideally, in work of this kind, the birds should be kept under observation throughout their waking hours, in order that we may have the full history of their behavior. Such prolonged watching is impracticable, particularly since during much of the day there is little pecking. Actually, observations were restricted to the time near feeding, when the birds were most likely to fight. Taken together with the greater number of triangles, the reversals indicate a less stable social order among these male birds than among their sisters.

For a time there was no completely dominant bird among the cockerels. BW, which stood highest in general, was pecked by BR, which ranked otherwise just below him. One day BR and Y started to fight, as they had done many times before, with BR winning. This time Y struck through to the eye, which closed as a result, and BR retreated. The injury was such that the tenderhearted observer thought that BR needed special treatment, and removed him to a hospital pen. The eye healed, and two weeks later the recovered bird was returned to the flock that he had almost dominated. In these two weeks of absence he had lost his social status entirely, and was pecked even by W, which had not been seen before to peck a fellow cockerel. The reason for his loss of position is not clear. He had been severely injured, he had lost a fight to an inferior, and he had been absent from the flock for fourteen days. For one or all of these reasons he had lost caste so completely that five days later he had to be removed from the flock, literally to save his life.

During the five days that BR was again with the flock, he avoided contacts with others as much as possible and spent a great deal of his time crowding under a low shelf on which the water dish was kept. In our experience, the lowest-ranking chicken in a flock tends to avoid social contacts as BR did after his fall from a superior position. Frequently the low-ranking birds show many objective signs of fear. They spend time in out-of-the-way places, feed after others have fed, and make their way around cautiously, apparently with an eye out to avoid contacts. The lowest-ranking birds may appear lean, and their plumage is somewhat more rumpled because they have less time to arrange it. Dominant birds, on the other hand, are characterized by a complete absence of signs of fear or of any attempt to avoid birds of lower rank. Some birds, usually those high in the peck order but not at the top of it, show few avoiding reactions to their superiors and when pecked apparently take it lightly and pass on.

The possible correlation between social rank of individuals and their survival value is relatively easy to investigate. The survival values of organized as compared with socially unorganized groups are more important and much harder to test effectively. The nearest approach we have been able to make has been to measure various effects produced when mature hens are allowed to establish and maintain a peck order in flocks with the same membership, as compared with other flocks in which the hen longest present is replaced daily, or every other day, by a total stranger.

Individuals in the unchanged organized flocks pecked each other less, ate more, maintained weight better, and laid more eggs than did their fellows in the flocks steadily undergoing reorganization. The indications are that the social organization in these flocks is of value not as an end in itself but because it tends to reduce fighting and other extremes of social tension.[13]

Chickens show some other interesting reactions that are related to their position in the social hierarchy to which they belong. Professor C. Murchison, a psychologist formerly at Clark University, has reported studies on the behavior of a flock of six cocks and five pullets.[94] In one series of experiments, pair after pair of the cocks were selected at random and placed at either end of a narrow runway behind glass doors that allowed them full sight of each other. When the glass doors were opened, the cocks ran toward each other. The point of meeting was proportional to the relative position of the two in the social scale, for the more dominant bird traveled farther than the subordinate one.

In our studies we have usually found that the birds higher in the social order had more social contacts than those that were at the bottom of the peck order. The correlation is not always exact, but to date we have found few exceptions to the rule that the bird lowest in the peck order has the fewest contacts. A quantitative difference, closely associated with social rank, may be found in the number of pecks delivered when there is no difference in the total contacts among the upper birds. In one study[10] in which four pens with five or six pullets in each were under observation, out of 4,400 pecks the ranking birds gave 1,800, the second in the lists gave 1,092, and so on in regularly declining numbers until those next to the bottom gave 136 and the birds that were lowest in their respective flocks gave none at all.

When several cocks and hens are in the same flock, high-ranking cocks may suppress the mating behavior of some subordinates and, on the other hand, may even allow others to push them away and replace them in the copulating position. High-ranking hens give the so-called mating invitation less frequently than do their low-ranking associates: that is, they crouch less frequently before a cock. They also are courted less and are mated with less often than are the hens below them in the social hierarchy.[11]

These are some of the known relationships existing among birds that have a relatively fixed group organization. Schjelderup-Ebbe,[119, 120] who has made observations on over fifty species of birds, including, besides the common chicken, a sparrow, various ducks, geese, pheasants, cockatoos, parrots, and the common caged canary, is convinced that despotism is one of the major biological principles; that whenever two birds are together invariably one is despot and the other subservient and both know it. He has said, "Despotism is the basic idea of the world, indissolubly bound up with all life and existence. On it rests the meaning of the struggle for existence." He applies this principle to interactions of men and of other animals and even to lifeless things. He says: "There is nothing that does not have a despot . . . usually a great number of despots. The storm is despot over the water; the lightning over the rock; water over the stone which it dissolves"; and he cites with approval the old German proverb that God is despot over the Devil.

This poetry of Schjelderup-Ebbe's is striking, but does it rightly interpret the facts? We have spent a considerable amount of time at Chicago, investigating the social order of various birds. We have not yet studied as many varieties of birds as Schjelderup-Ebbe, and we have no experience to report about the relation between God and the Devil. Of the birds we have studied, only the flocks of white-throated sparrows approach the common chickens in the fixity of their social hierarchies, and they do not equal it. The common pigeon, the ring dove, the common canary, and the parakeet show a less rigid type of social organization. I can illustrate it by explaining the situation as we have found it among common pigeons.[90]

The observations were made on a group of fourteen white king pigeons, half of which were male and half female. Their social order was observed in sex-segregated flocks until, after a month, it seemed to be fairly stable; then the flocks were combined, and after a month, during which five of the seven possible pairs mated, the sexes were again segregated for twenty-eight days of further study. The results are essentially similar both for the males and the females for the period when the sexes were separate, so that I shall follow only the reactions of the female flock. The essential facts can be described with the aid of the diagrams in Figure 41. These show the social interactions between the females lowest in the social order.

Let us examine A with some care. It charts the relationships of the five birds that were lowest in the premating flock. All these were dominated in the main by BY and BB. The figures show that BR was seen to peck GW ten times and was pecked by GW, and retreated

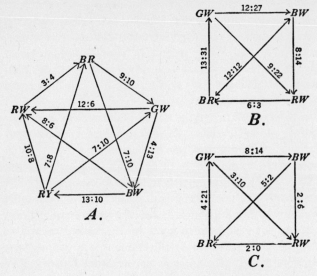

FIGURE 41. *In flocks of pigeons the organization is one of peck dominance rather than of peck right. The pigeons highest in the social order are omitted from these diagrams. A, the pre-mating flock; B, the entire period of observation; C, the post-mating flock.*

from her nine times. GW pecked BW thirteen times, but lost in four encounters. BR won ten and lost seven of its observed contacts with BW, which won thirteen and lost ten with RY. RY in turn was practically even (eight to seven) with BR and slightly ahead in its relations with GW and RW. I do not intend to suggest that most of these differences are important; in fact, that is the point. With flocks which are organized as are these pigeons, it frequently becomes difficult to decide which bird stands higher in the social order.

It is important to note that in none of these cases—in fact, in only one of all the different reaction pairs whose behavior is summarized in these charts—was there an absolute dominance of one bird by the other, and then only two contact reactions were seen. When all contacts throughout the whole period of observation are considered, there was at least one time for each of the contact pairs when the bird that usually lost out dominated the contact reaction.

In B, which shows all the reactions during pre- and post-mating, and in C, which records the contacts for the post-mating season only, the four birds represented by the diagrams were dominated by three others, RY, BY, and BB. It is worth emphasizing that with these birds an absolute despotism was not established. Even RY, which more than any other bird dominated the post-mating flock, lost contact reactions to each of the others except to RW, which was lowest of all. While it was winning 329 reactions it lost 58, and each of the other females, RW excepted, dominated it at least three times in the post-mating observations.

The picture that emerges is one of a flock which is organized into a social hierarchy, but one which is not so hard and fast as that found with chickens. In the long run one becomes fairly sure which bird in each of the groups will dominate in the larger number of their contacts, but the result of the next meeting between two individuals is not to be known with certainty until it has taken place. Within the same hour and even within a few minutes reversals in dominance may take place without anything unusual in the circumstances.

Putting the matter somewhat facetiously, chickens appear to have developed the sort of "line organization" characteristic of a military system or a fascist state, whereas these pigeons, together with the ring doves, canaries, and parakeets, are more democratic. The social hierarchy among chickens is based on an almost absolute peck right

which smacks strongly of the despotism of which Schjelderup-Ebbe writes, whereas these other birds have an
organization based on peck dominance rather than on
absolute peck right.

With such birds social position is not fixed once and for
all. Consider the case of RY among pigeons. When results
were first put together at the end of two weeks of observation, RY was at the bottom of the flock, a position
that it retained for twelve more days. Then something
began to happen. What it was, I wish I knew. RY began
to go up in her social world. After six days she ranked a
shaky third, clearly dominated on the average by BY and
BB.

Then the pigeons were allowed to mate. During the
mating period, BY, which was top bird in the pre-mating
flock of females, and RY did not pair off with any of the
males. Again I do not know why. After the experiment
was finished, RY was carefully autopsied and we could
find no evidence of anything physically abnormal. When
the sexes were again segregated, RY was the top-ranking
bird among her fellow females, and remained so. She was
seen to have 101 contacts with BY, the former alpha bird,
and to win 83 of them; she had 77 observed contacts with
BB, which had formerly been second from the top, and
defeated her 53 times. In the pre-mating period, RY lost
two combats for each that she won; in the post-mating
flock, she won five contacts for each that she lost.

This raises in a rather dramatic fashion questions as to
what qualities make for a dominant bird. This problem is
not yet solved. With these birds, social rank is in part a
matter of seniority. Mature chickens usually dominate
immature ones and maintain their dominance long after
the former youngsters have become fully mature and
possibly physically able to displace the senior members.
This is good evidence that memory of former defeats

plays a role in maintaining the social order once it is established. When chickens strange to each other are put together for the first time, dominance usually goes to the bird with superior fighting or bluffing ability. Maturity, strength, courage, pugnacity, and health all seem essential qualities making for dominance among chickens. Luck in combat also seems to play a part when one considers the numerous triangle situations that have been discovered. Since cockerels have certain of these qualities more than pullets, a male bird, if present, dominates a flock of hens.

There seems to be little if any correlation between greater weight and position in the peck order. The location of the combat seems to be important. Schjelderup-Ebbe found that chickens in their home yard win more combats than strangers to that yard; and Dr. Shoemaker has reported that, with canaries, each bird becomes dominant in the region near its nest.[124] We found some years ago that with pigeons one might be dominant on the ground about the feed pan and another have first rank at the entrance to the roosts.[90]

With chickens, as I have said, the larger, stronger, more pugnacious males usually dominate the females. This is said to be generally true in species in which the male is larger or more showy than the female. With the parakeets,[15] whose social order in many ways resembles that of pigeons, the females are dominant over the males except in the breeding season. While breeding and nesting are in progress, positions are reversed, and a previously hen-pecked male may drive his usually dominant mate back onto the nest when she attempts to leave it. The sexes in these parakeets can be told apart only by slight differences in color.

When hens are giving the brooding reaction or are caring for small chickens, they become less submissive to other hens. Some of the other birds, whose social ranking has been investigated, move up and down in the social scale according to the phase of the breeding and nesting cycle that they are in at the time.

It has been reported that with hens those high in the peck order have a higher I.Q. than their more lowly placed flock mates.[80] The I.Q. was measured in this case by placing grains of corn in a line on the floor with every second grain securely fastened down, and finding the speed and accuracy with which the fowls would learn to peck at the loose grains only.

We have had as yet only the most casual personal contact with this problem so far as chickens are concerned. With the parakeets, Masure and I could find no evidence of a positive correlation between any aspect of ability to learn a maze and social rank.

From this summary it is evident that in spite of a great deal of study we do not know all the factors that determine the position of birds in their social order. Known factors include: strength, state of molt, presence of male hormone (males usually dominate females), size, weight, maturity, fighting experience and skill, stamina, sensitivity, fatigue, previous social rank, resemblance of the opponent to a despot or subordinate in the home flock, strangeness of territory, presence of known associates, and such considerations as chance blows and other unpredictable happenings.[11]

Some of the complications in determining the factors that make for dominance are shown by the preliminary summary that Dr. Shoemaker made from his studies on the social hierarchy in canaries. The space available for the caged flock is a matter of importance. When confined in relatively small space, the social order becomes more simple and definite and there is no complication over the question of territorial rights. With more space—as, for example, in a large flight cage—individual territories tend to become established in which the particular bird is supreme even though it ranks low in the neutral ground around the bath bowls, the feeding places, or regions where nesting material is stored.

When canaries are allowed to mate and small nesting cages are supplied around the walls of the flight cage, each individual male is master in its own nest cage and controls more or less territory around the cage entrance. Under these conditions even the birds lowest in the social order dominate in some restricted space about their nest.

In general these canaries show more pecking among the males than among the females, and during the nesting period the female does little to defend the nest territory; that is the work of her mate. In this home territory the social dominance of the male over his fellow males is not steady but varies with different phases of the breeding cycle. During the processes of nest-building, egg-laying, and incubation, the male tends to become more dominant. This is shown by an increase in the size of the territory about the nest that he dominates, and by the fact that when in neutral territory he tends to win more of his pair contacts. During the rest of the cycle the male tends to lose dominance as measured by both these criteria.

It is worth noting that in the course of these pulsations in dominance the male may not actually move up in the social scale as determined by the number of birds that he fully dominates. He may win more of his individual pair contacts without actually upsetting the usual trend. The same bird may show fluctuations in dominance during the day. Thus one male regularly dominated less territory in the evening than he did in the morning. This may well be a matter of stamina.

In some cases the relation between the sexes in these canaries hinges on another complication. For example, female number 15, mated with male number 55, which stood about midway in the social order among the males: 55 dominated the other females and all the other males dominated over 15. However, of thirty-three observed contacts between 15 and 55, the male lost all but one! The male parakeet will drive back to her nest the female who has left it, but 55, like other male canaries, coaxed his mate back to her nest with offers of food.

Until studies are further advanced, we cannot be sure how many of these complications that Dr. Shoemaker has recorded for the canaries are found elsewhere even among birds. It seems reasonable to suppose, however, that the social hierarchy is rarely as simple in its organization as a mere listing of the social ranking seems to indicate.

With all these birds, high rank in the social order of the flock means much greater freedom of action, more ready access to food, and a generally less strained style of living. It is hard to say whether in nature it means more than this, although it seems probable that in times of food shortage, or other phases of environmental stress, the ranking birds, which have the first opportunity at food, might readily fare better than those low in the social scale. Fortunately, enough observations have been made in nature so we know that with some species the peck order, which has been most studied in restricted cages and pens, does occur in the wild.

The alpha bird in a penned flock of chickens does not necessarily lead in foraging expeditions when the flock has more space. Fischel reports that when hens of known peck order are released to forage in an orchard the dominant and near-dominant birds may or may not be at the apex of the foraging flock.[52] Usually the leadership changes from time to time; moreover, the leading bird seems always more or less dependent upon her followers.

If she gets too far out ahead the leader turns back and re-joins the flock or waits for them to catch up. Similar hesi-tation by the leader when it has advanced some distance in front of its followers has been observed among other animals, notably among ants and men.

This problem of leadership among birds is related to, but not identical with, position in the social order. There are many aspects of the problem into which we cannot go at present, pending a closer and more revealing study than has appeared as yet of the qualities that make for leadership.

With some herds or hordes of mammals leadership rests with an old and experienced female.[20] In such herds the females and young frequently make up the more stable part of the social group, to which males attach themselves during the mating season. With other mammals the male is the leader, and sometimes a jealous one, who drives other males out of the herd; although in some cases sev-eral males are tolerated.[3]

Leadership does not always go to the faster or stronger animal; in fact, the position of being out in front of the flock may not mean real leadership. An interesting ex-ample of such pseudo-leadership has been recorded for a mixed group of shore birds observed by Mr. J. T. Nichols, of the American Museum of Natural History.[96] He found a mixed flock of such birds, composed of two young dowitchers, a dozen black-bellied plovers, and a single golden plover. Under these conditions certain of the birds could readily be distinguished from the others. When the flock was flushed, the flight of the golden plover was comparatively rapid and this bird was soon ahead of all the rest. The dowitchers were slow and tended to fall behind, and when this happened the black-bellied plover wheeled. This affected both the apparent leader, the golden plover, and the lagging dowitchers. The former,

finding itself alone without followers, rose above the
flock, took the new direction and dived down with a few
swift wing beats, again the apparent leader of them all.
The slower dowitchers took the chord of the arc made
by the wheeling flock and so caught up with and again
became an integral part of the flying group. Soon again
the slow dowitchers lagged and the whole performance
was repeated.

These observations do not reveal the stimulus releasing the wheeling mechanism of the main flock. The simplest explanation, that the leader, finding himself out
alone in front, starts to turn and so gives a stimulus to
the keen-sighted remainder so that they also shift direction almost instantaneously, does not hold in this
mixed flock, for the observations indicate clearly that the
apparent leader, the golden plover, was following along in
front of the main flock as much as the slow dowitchers
were following along behind it.

Neither does this simple-leadership sort of explanation
fit the facts as observed among wheeling flocks of other
shore birds or of pigeons. In such flocks the stimulus to
turn frequently seems to originate in one of the flanks,
and it spreads from that point rapidly through the flock.
Here again the apparent leaders may not be the actual
ones. It is possible, though we are not yet sure of it, that
in such flocks made of birds that we cannot tell apart, the
faster individuals also may dive through the flock to the
foremost position, taking their direction from the whole
flock.

However the signal for turning originates, the wheeling takes place so rapidly that mythical explanations are
still being advanced. I have a small book written on the
subject by an English author, called, *Thought-transference (or What) in Birds?* [122] The title correctly summarizes the contents of the book.

I would not have you conclude from my repeated emphasis on the absence of definite leadership in these flocks of birds, and on the presence of a pseudo-leadership when the flock is really determining the direction that is taken by the bird in front, that there is no real leadership among other animals and among men. And I must make it clear that here I am speaking of real leadership and not of a peck order, which, as is true with social position in human society, does not imply leadership at all. Such a position could not be successfully maintained by a person trained in science rather than in dialectics. But apparently, at least among so-called lower animals, the leader is frequently as dependent upon his followers as they are on him, and sometimes even more so. A similar situation occurs in human affairs often enough and under such a variety of situations that the relationship deserves more careful consideration than it usually receives when problems of leadership are discussed.

Although those of us who have been engaged in these studies have probably never been wholly unaware of the possibility of amusing cross-references to man, I must insist that our motivation has not been that of making an oblique attack upon human social relations. Rather, we have found problems concerned with the social organization of birds and other animals interesting and important on their own account.

We have, of course, a feeling that different animals have much in common in group psychology and in sociology, as well as in more distinctly physiological processes. It is the viewpoint of general physiology that we cannot understand the working and the possibilities of the human nervous system, for example, without study of the functioning of the nervous systems of many other kinds of animals. Similarly well-integrated information has been compiled concerning general and comparative psychology. From the same point of view some of us have been trying to develop a general sociology, which even in its present imperfect state allows human social reactions to be viewed in part as the peculiar human development of social tendencies that also have their peculiar developments among insects, birds, fish, mice, and monkeys: that is, among social animals generally.

Keeping this point of view, and with our background of studies of social organization, it is worth while to turn to a short consideration of the actual application of similarly objective studies in certain human groups. I pass over the possibilities of studying the peck-order in women's clubs, faculty groups, families, or churches, to call your attention to some studies that have been published dealing with the social interactions of the Dionne quintuplets, since these will serve to throw light on a number of interesting points.[29]

In all questions of dominance in the group or of other forms of social inequality, we come immediately and continually upon the question of the extent to which these observed social differences are a matter of heredity and to what extent they follow differences in training or other environmental impacts. This is the old nature-nurture problem, other aspects of which have been discussed for years.

Driven by many different kinds of evidence, biologists have come to the conclusion that all men are not born equal. Applying this to social affairs we have the general assumption that many of the observed differences in social position are a result of the inherited differences depending on the vagaries of biparental inheritance and more remotely on mutations of one kind or another.

Fortunately we have in the case of the Dionne quintuplets a natural experiment that deserves much attention. Detailed biological studies which appeared late in 1937 confirmed the general assumption that these much-discussed children are an identical set of sisters. Biologically this means that all of them have come from one ovum that was fertilized by one spermatozoan. Soon after fertilization the early cleavage cells separated and produced five embryos, each with identical heredity. I shall not give the details of the evidence on which this conclusion is based. In addition to looking so much alike that only their regular attendants can tell them apart with any degree of sureness, there are similarities in finger and palm prints, in toe and sole prints, and in other anatomic details that point conclusively toward a common identical heredity.[29]

A group of investigators from the University of Toronto studied the social reactions of the quintuplets and reported observations from the twelfth to the thirty-sixth month of their age. At first the children were placed together in a play pen by pairs to observe their interactions; from the twenty-second to the thirty-sixth month they were observed as a group.

The available records do not allow an exact comparison with the peck order I have described for various birds. The observers were interested in recording and analyzing the following bits of behavior:

1. Total contact reactions.

2. Reactions of one child toward another, which they call *to* reactions. A *to* reaction by one child will be a *from* reaction for the child receiving the attention.

3. Whether the reactions are initiated or are response bits of behavior. An illustration will help to make this clear. If A pushes Y, it is regarded as an initiated *to* reaction by A, while Y is credited with a *from* reaction. If Y pushes back, then this is a response *to* reaction for Y and a *from* reaction for A.

4. They also record which child watched which one.

I shall not use all these distinctions, for my points can be made accurately with only some of them.

In Figure 42, certain reactions are summarized in the top row for the entire period from the twenty-second to the thirty-sixth month, and in the lower row the same reactions are summarized for the last four months of the study, from the thirty-second to the thirty-sixth month. The left-hand diagrams give the total contact reactions during these respective periods. The center diagrams show the total *to* reactions and those on the right give the initiated *to* reactions.

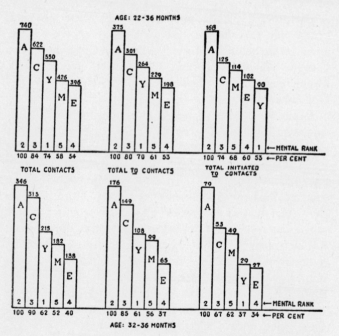

FIGURE 42. *The Dionne quintuplets show evidence of a social organization among themselves.*

Let us examine the upper left figure. A had a total of 740 observed contacts directed to and received from her sisters. This is taken as 100 per cent. C had similarly 622 contacts, which were 84 per cent of A's, and so on, with Y third, M fourth, and E fifth. The order in total number of contacts, then, is A, C, Y, M, E. This same order holds for total *to* contacts and for both total and *to* contacts in the thirty-two to thirty-six months' period. The diagrams on the right show that A initiated the most *to* contacts, and that C was next. Beyond that the order varies. For the whole period of observation (upper row) it is M, E, Y, and for the last period (lower row) it stands M, Y, E.

The other available data do not always give this same order, but enough has been presented to show that, among these children identical in heredity and almost so in postnatal environment, there are social differences which can be recognized by the behavior of the children toward each other.

As the figures giving mental rank indicate, the correlation with intelligence is by no means perfect. Neither is the correlation with size. Y, the largest, and said in some ways to be the most mature of the five, ranks in the tests shown here from third to fifth. And while M, the smallest, ranks low, she is not the lowest, and other data show that in the percentage of her contacts that were self-initiated *to* reactions she ranks first of all these sisters.

These observed differences raise an interesting question: If heredity has been the same and the environment constant, how did the differences creep in? It is possible that there are unobserved, unrecognized differences either in the handling of the children, in their early contacts with each other, or in their impacts with their physical environment that may have been cumulative enough to produce these social differences. It is also possible, as

Professor H. H. Newman suggests, that the differences are environmental after all. We must remember that from the standpoint of A, C, Y, M, and E, their environmental relations began long before birth, and though the care given them since birth may have been practically identical in each case, it may not have been possible to erase environmental conditions impressed upon them during their seven critical months of intra-uterine life.

Whatever the reason, we have come to an interesting, and, I think, important conclusion, which is that animals with exactly the same heredity may still develop, even at an early age, graded social differences showing that one is not exactly equal to the other. We have indications that the same principle holds among birds, but even if present indications are finally borne out, the experiment will not be as elegant, in the strictly scientific sense, as are these observations on the Dionne quintuplets.

Finally, by way of review, there exists among flocks of birds, even though they may be identical to the human eye, a graded series of reactions within the flock that allow observers to rank the birds in the order of their social dominance. This social order may be relatively hard and fast, as with hens, or more loosely organized on a give-and-take basis, as among pigeons and canaries. The factors underlying the social order in these birds are complicated and include such personal traits as age, pugnacity, sex in general and the reproductive cycle in particular, as well as such environmental factors as size of available space and the possibilities of establishing special territories. High position in the social order does not necessarily coincide with group leadership, although at times it does. The survival value of high position in the social hierarchy has not been demonstrated, but there are many reasons for suspecting that it may be felt in times of famine or during other periods of environmental stress.

The problems related to leadership, although mentioned, were not discussed exhaustively. Emphasis was laid on the importance to the leader of his followers, and on the existence of a pseudo-leadership in which the animal in front is taking direction from his apparent followers.

With the Dionne quintuplets it was demonstrated that social differences exist even with children that have identical heredity, and a theory of environmental differences was favored as an explanation.

In conclusion, the social organization observed in birds and other animals reminds one almost constantly of certain types of human social situations. The dominance-subordination relations of people are at times readily observed; at other times they are obscured by other social responses. When present in man, patterns of domination may be expressed in many more ways than in birds or mice. It may well be that the social hierarchies of chickens, canaries, and men have much in common. Without taking the comparison too seriously, the fact that chickens, for example, have a relatively simple system of despotism may help explain, though it does not justify, the appearance of a similar social organization in man. Other types of social organization also exist among the other animals, and man need develop only that best suited to his unique situation.

Social Transitions.
Chapter I X

When does an animal group become truly social? This question has already arisen in preceding chapters and is difficult for a thoughtful biologist to answer with confidence.

One school, now happily small, regards society as beginning when animals first display a social instinct.[20] By this they probably mean that social animals have inherited a behavior pattern that causes them to live together with others of their kind in more or less closely cooperative units. Others consider that animals are social when they carry on group life in which there is clear evidence of a division of labor.[47] There is also the frequent suggestion that only those animals are truly social whose behavior is an extension, directly or indirectly, of familial behavior.[131]

For myself, I regard those groups in which animals confer distinct survival values upon each other as being at least partly social; this is the conception that has most often appeared in these pages.[3] And from a still different point of view, those who would stretch the idea of social living rather widely would say, as I have indicated in Chapter VII, that when animals behave differently in the presence of others than they would if alone, they are to that extent social.[127]

These ideas concerning what constitutes a proper definition of animal societies, although not necessarily mutually exclusive, are sufficiently different to raise difficulties when one tries to examine critically the useful general concept of social life; it will be profitable to study some of them separately.

As to the first definition, that social life must be limited to those animals that possess a social instinct—an inherited behavior pattern—it is difficult to demonstrate beyond reasonable doubt that many patterns of social behavior are in fact inherited. Is the tendency of many fishes to form closely knit schools inherited or an early-conditioned bit of behavior? There is some evidence that it is inherited, but we are not yet sure. And even if it were granted that such schooling tendencies are innate, it would not necessarily follow that they are instinctive. There are different degrees of complication of inherited behavior patterns, from the relatively simple reflex action of an unborn embryo to the complex mating behavior shown, for example, by some insects and by rats. The exact determination of the place in this line of increasing complexity at which an action ceases to be a simple reflex and becomes a more elaborate reflex-like action of the entire organism, or the point at which the latter becomes sufficiently complex to be called an instinct, has never been made. That is, we do not know just how far down in developing patterns instinctive behavior extends.

There is the added complication that the word "instinct" has been loosely used. The most workable definition that I have arrived at is a modification of an older one of Wheeler's: An instinct is a complicated reaction that an animal gives when it reacts as a whole, and as a representative of a species rather than as an individual, which is not improved by experience, and has an end or purpose of which the animal cannot be aware. Too frequently the word has been applied to any bit of behavior whose origin and motivation the observer did not understand, with the unfortunate paradoxical implications that thereby the action was explained and at the same time could not be further explained. As a result of this uncritical usage many careful workers disapprove of employing the word under any conditions, and particularly in the field of social activities.

In recent years some students of social life have attempted to avoid the term "social instinct," while employing the same fundamental idea under the thin disguise of "social appetite,"[132] "social drive," or "group interattraction,"[110] all of which are apparently understood as having been inherited. These contributions to a more picturesque language do not necessarily advance our understanding of social behavior.

Still others sincerely believe that behavior patterns are not inherited, which seems to me a clearly untenable position. But however strong my belief in the actual inheritance of social behavior I do not consider it helpful to make the possession of such an inheritance the major criterion of social living; it is not a practical working test as to what constitutes social life.

If division of labor be used as a touchstone the same type of difficulty arises. We do not know how to determine when such a division becomes sufficiently general to merit being called a social attribute in the stricter sense in which we are now using the term. For example, there is a division of labor that is associated with sex and that is almost as extensive as sex itself. When does this particular division of labor cease to be merely an expression of sex and become social in the commonly accepted use of the word?

The mention of sex brings up again another important definition of social life among animals, which has already been listed. This states that only those groups are truly social that have grown out of the persistence of sexual pairs or groups and, more especially, that have developed from partial or completely familial relations. This point of view has been touched upon with some sympathy in an earlier chapter. An important relationship underlies this definition: many highly organized social groups do develop from the continuation and extension of family ties. But though this condition has given rise to many of the better developed social units, care must be taken not to regard its presence as the essential difference between the social and the subsocial. As Professor C.M. Child[35] has suggested, boys' gangs, girls' cliques, and men's and women's clubs present difficulties to one who wishes to define all societies as extensions of familial relationships. It is quite possible to regard such social phenomena as expressions of aspects of the social urge that have developed independently of paternal or fraternal interactions. There are counterparts of these human groups among other animals, as well as counterparts of the extensions of family life. The overnight aggregations of male robins, the long-continuing stag parties of male deer outside the short rutting season,[43] the flocks of mixed species of birds common in tropical regions (Beebe tells of one made up of twenty-eight individuals representing twenty-three species),[28] schools of fishes, and the swarms of animals spoken of in the third chapter—all of these instances test and stretch in varied ways the idea that only those continuing aggregations of animals that grow out of sexual and familial interrelations are truly social.

Inherited behavior patterns, which are the forerunners of instincts, and sexual differences extend down to the protozoa; so do continuing family groups, especially in the form of structurally connected colonial organisms. Group survival values are present in groups of organisms in which sex has not yet evolved, as well as among those in which sex is elaborately developed. In the light of such considerations it becomes exceedingly difficult to establish any one line above which life is to be regarded as truly social and below which we have only differing degrees of subsocial relations. Here, as happens so frequently in biology, we are confronted with a gradual development of real differences without being able to put a finger with surety on any one clearly defined break in the continuity. The slow accumulation of more and more social tendencies leads finally by small steps to something that is apparently different. If we disregard the intermediate stages the difference may appear pronounced, but if we focus on these intermediates it will be only for the sake of convenience that we interrupt the connecting chain of events at some comparatively conspicuous link and arbitrarily make this the dividing point, when one is needed, between the more and the less social. It must be recognized that any such division is a matter of convenience rather than a natural break in the development from mass or simple group behavior to highly evolved social life.

For our purpose in the present account it is sufficient to recognize that the well-integrated social systems of man and other mammals, of bird flocks, and of insect colonies exhibit among them the highest expressions of social abilities that have evolved. In the range of social development shown in these animals we find attributes that are truly social in the most exclusive use of the word. But these highest expressions of social living have their roots in tendencies that in the form of unconscious cooperation accompany animal aggregations extending throughout the whole animal world, as well as to some extent among plants. Conceding then the difficulties in the way of making any exact definition of social behavior, I wish to present some of the social implications of mass physiology, particularly among well-integrated societies of animals.

One of the characteristics of social life among the insects is the presence of castes[133] that perform different functions within the colony. With many social insects the division of labor has developed to such an extent that the animals' bodies are more or less structurally appropriate to their principal tasks. The reproductive female has a greatly enlarged abdomen; the soldier grows up to possess large jaws and heavy armor or other protective and attacking devices; a worker may be large or small or medium in size, according as its size will best suit for some of the varied tasks necessary for the life of the whole colony. The situation is greatly different from that among human social castes, where a member of the aristocracy may be as husky of body and as empty of mind as the most menial of the working caste.

The only physically distinct castes to be found in man and the higher vertebrates are those associated with sex. In sexual forms there is always a division of labor with regard to the primary sexual functions except in those rare cases, usually low in the evolutionary scale, which are at one time both male and female. With many, aside from producing eggs rather than sperm, it is difficult to find a division of labor or of appearance between the sexes. With others, particularly among the more specialized animals, there are differences in sexual behavior and responsibilities that are associated with the more fundamental distinctions of sex. Frequently, as in man, these differences have developed into fairly distinct behavior patterns for the two sexes, until each sex is practically a distinct caste, almost in the sense used in discussing castes among the social insects.

Sex is usually determined by differences in heredity associated with the combination of chromosomes[42] and of the bearers of heredity (genes) that are found in the sperm and egg whose union gives rise to the new individual. Such determinations occur at the time of fertilization and sex is normally unaltered thereafter.

Exceptions occur demonstrating that for certain animals this normal means of sex determination can be overruled by environmental differences. Many of these cases are interesting and significant, but their full consideration here would draw us off the main thread of our present discussion. We shall follow only those instances in which changes in sex are associated with the near-by presence of other individuals, considering here two widely differing cases that have been carefully investigated in recent years.

Professor W.R. Coe,[39] of Yale and California, has spent much of his time studying the sex ratios and sexual changes in oysters, clams, marine snails, and other related forms. In many of these mollusks he has found that the sex ratios vary greatly in different environments, and he has reached the conclusion that frequently among these animals the expression of an innate sexual tendency may be in part suppressed or stimulated, as the case may be, by the environment in which any given animal is living.

A pertinent case is that of a set of marine snails of the genus *Crepidula*. Three of these "boat-shell" snails are common animals in the coastal waters of southern New England. Their sexual history follows similar outlines. After a juvenile period that is essentially asexual, the growing *Crepidula* first becomes a male and later, sometimes only at long last, transforms into a female. A typical species to follow through this transformation is *Crepidula fornicata*.

When young, these animals move about, but as they become older and larger they settle down in one place on a wharf piling, a rock, or another shell. If the larger, older animals are broken loose, the soft parts are usually destroyed by some predator before they can reattach themselves, leaving behind the relatively heavy shell. Frequently they form chains of individuals, of which a simple example is shown in Figure 43. The large, bottom snail is dead.

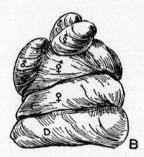

FIGURE 43. Crepidula fornicata. A. *A basal female is attached to a dead shell* (D); *two individuals are in transition stages and there is one male at the apex; three motile supplementary males are in mating position on the lower transition individual.* B. *Same group from the left side.* (*From Coe.*)

Attached to its shell is a large female that in summer actively produces eggs. Above her are two individuals that are undergoing transformation from male to female. Scattered about over these are four smaller snails, which are still functional males and which can and do move about. Each male has a long slender penis by means of which he transfers sperm from his body to an appropriate receptacle in the body of the female. Several males may participate in the insemination of a single female.

The growth of these snails is fairly rapid. A young snail hatched out early in the summer may, before autumn, become a functional male about 16 mm. long, which is about two fifths the size of a fully adult female; during the following year he will probably transform into a female.

The relationships that Dr. Coe observed at Woods Hole may be summarized briefly. Some two hundred young males were taken from their normal surroundings and placed in separate containers in the laboratory. Two months later only 11 per cent were still functional males; 15 per cent had transformed completely into functional females and the other 74 per cent were on their way in that direction. Random collections of snails of similar sizes that had been left alone in their natural associations showed that 85 per cent were still functional males and only 3 per cent had fully changed into females.

Coe summarizes his work with this and the other Crepidulas as follows: "There is no doubt but that in each of these three species of *Crepidula* stable environmental conditions tend to prolong the male phase of these individuals that are suitably mated and sedentary." These points are further illustrated in his diagram, a part of which is reproduced in Figure 44.

There is evidence from the earlier work of other observers,[61, 98] which these studies by Coe do not entirely

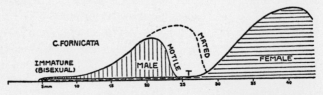

FIGURE 44. *As* Crepidula fornicata *gets older and larger it passes successively from the sexually immature through the male on into a final female stage. Mated males retain that stage longer than if actively motile. (From Coe.)*

replace, that association with a female is important for the full realization of the male condition as well as for its prolongation. With these snails the tendency to become first a male and later a female is probably determined by heredity, although the hereditary mechanism which promotes such a shift is at present unknown. The point of interest for this discussion is that the association with others, especially among mated males, tends to postpone transformation to the opposite sex.

Some cases are known in which the presence of other animals of the same species determines the sex. One of the most thoroughly studied is that of the worm *Bonellia*,[25] in which the sexually undifferentiated larva does not, in nature, become the small parasitic male unless it is associated with the large female.

Among certain nematode worms that are parasitic in insects, if few eggs are introduced into, for example, grasshoppers,[3] most of the resulting nematode parasites are females; but if many eggs are fed, the nematodes that hatch are almost all males. The results are not to be ascribed to a differential death rate, for approximately 75 per cent of the eggs develop in both cases.

In *Crepidula*, *Bonellia*, and nematodes, both males and females are always present in a population, though in different ratios. In cladocerans, however, of which *Daphnia* is an example, the species may be carried along for many generations by the females alone. They produce eggs that do not require fertilization but develop directly into females that again produce other females like themselves. In these cladocerans the race is usually made up of females alone, but at times there is an outbreak of sexuality; males and sexual females appear and the fertilized eggs resulting from their union are more resistant to adverse conditions than those that are ordinarily produced and that require no fertilization. These resistant eggs enable the species to survive times of environmental stress, such as winter's ice or the drying up of the ponds in which these small crustaceans live.

In one species of *Moina*,[4] which has been much studied by the biologists at Brown University, crowding of the females is an effective method of bringing on the outbreak of males and sexual females, so that overcrowding may be rated as a time of environmental stress. Either by the shortage of food, by the accumulation of waste products, or from some other cause, the association of many female cladocerans together results in the production of eggs that have a different prospective potency from those the same females, uncrowded, would produce; and sexual males and females are the result.

It is evident from these varying examples that even the fundamental matter of sex, with the caste-like divisions of labor that result from two sexes, may be determined by the close association of animals of the same species. There is some reason for suggesting as in Chapter IV that sex itself may have grown originally out of mutual acceleration in division rates when two or more primitive organisms were in close contact in small space. The whole matter of sex may hark back to some of the basic aspects of mass physiology that were set forth earlier in this book.

Sex in its different aspects plays a highly important role in the social affairs of animals. It is interesting to find that this fundamental cleavage through so much of animal life can at times be controlled by group relationships. Such considerations serve again to emphasize the difficulty of drawing a hard and fast line, or even a fairly distinct band between social and subsocial living.

One phase of the social implications of sex has escaped general comment. I first heard it mentioned by the late Professor Wheeler.[133] Apparently when there is a social difference between the sexes it is the females that are the more and the males the less social; and the few striking exceptions only confirm the rule.

Among the social ants, bees, and wasps the normal affairs of the colony are carried on by the females. They produce males only when males are needed to fecundate the young virgin females at the time of their nuptial flight. The males contribute nothing to the protection, feeding, or housing of the colony; after their one sexual activity they die or are killed off, and the females that are lucky enough to secure a good nesting site carrry on with their female offspring until sexual repoduction again becomes the order of the day (Figure 45).

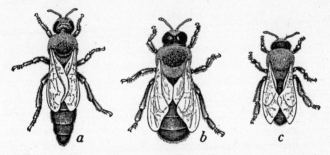

FIGURE 45. *Castes of the common honeybee: a, queen; b, male (drone); c, worker. (After Phillips.)*

With many of the herds of mammals, the main duties of communal life are borne by the females. They protect and rear the young and herd together to protect each other. The males keep to themselves except during the relatively brief period of the sexual rut. Even when they join the main herds, as in the case of the Scottish red deer, frequently the males do not fuse with the others. When danger appears during the rut, the stags make off and rejoin the females when the danger is past. After a male is sexually spent, frequently before the close of the breeding season, he withdraws, and the spent males form stag parties that are distinctly less social than the bands of females.

In commenting on the relative sociability of the sexes among red deer, Darling[43] says: "Matriarchy makes for gregariousness and family cohesion. The patriarchal group (among deer) can never be large, for however attentively the male may care for his group he is never selfless. Sexual jealousy is always ready to impinge on social relations leading to gregariousness. . . . I contend that the matriarchal system in animal life, being selfless, is a move toward the development of an ethical system."

The flocks of male birds whose social organization we have studied in Chapter VIII are more combative than the females. The human male writes the great poems, builds the great bridges, performs the outstanding scientific research; but he is also the criminal, the war-maker, the disturber of the peace. It is the human female that is the highly social force with our species, and in this we are again similar to the others mentioned.

Among the social animals only the termites have fully socialized males; with them the male reproductives consort with the female throughout life. Half the soldiers are males and the other half are females, and so with the workers. Termites are lowly insects, but in this one trait they lead the world. No one knows how the socialization of male termites was brought about, and even if we should learn their secret it probably could not be applied directly to human affairs.

When we turn from the far-reaching division of most animals into two sexual castes to explore the origin of the more specialized castes of insects, we find two different essential kinds, the reproductives and the sterile types. With bees, ants, and wasps, for example, the usual reproductive females can produce eggs that develop without being fertilized by a spermatozoan. Such eggs always give rise to males. From the store of sperm that she received in the nuptial flight the same female can allow her eggs to be fertilized; such fertilized eggs become females.

We have seen the comparative unimportance of the males. Although the active colony is usually composed of females only, these may be quite different in appearance and function. Typically there are the reproductive females and the sterile ones. Among the ants the sterile females are divided into the protective soldiers, whose main function is to protect the colony from the attack of other species of animals, and the workers proper. The ant workers are subdivided on the basis of size (Figure 46).

Professor Wheeler made the study of these social insects, particularly the ants, his life work. In a small book, published in 1937, after his death, he reaffirmed his belief that ants and bees have evolved from ancestral wasps and that each has developed the caste system independently.[136]

With bees and wasps, whether a given fertilized egg is to produce a worker or a sexual "queen," better called a reproductive female, depends upon the treatment and food given to the grub that hatches from the egg. If she receives plenty of food and is given space in which to grow, she becomes fully matured sexually; if fed less and kept more crowded, she becomes an incomplete female and is known as a worker. Apparently the fundamental difference can be brought about only by the treatment that the developing grub receives after hatching, and it is not a matter of heredity. Just how the workers are stimulated to give one or more grubs the treatment that will allow them to develop their full reproductive capacities is not fully known. If, however, the queen bee dies or is removed from the colony, workers will start enlarg-

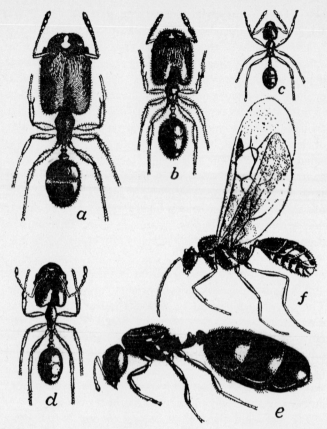

FIGURE 46. *Some ant castes: a, soldier; b, form intermediate between soldier and worker; c, worker; d, form intermediate between soldier and worker; e, queen that has shed her wings; f, winged male. (After Wheeler.)*

ing one or more of the cells that contain developing grubs, change their care and feeding and so allow them to transform into fertile reproductives. Perhaps they are kept from doing so when a queen is present by something like a social hormone, which there is good reason for thinking is produced by the even more social termites.

The mechanism that results in caste formation among ants need not be the same as that in wasps and bees, since it is generally conceded that they had a separate social evolution. For years two theories have been promoted as to how ant castes came into being. One group of students thought that ant castes were determined as were those of bees and wasps, by care and food; another group was equally sure that the differences were hereditary. After confessedly wavering between the two views in his long study of ants, Professor Wheeler in his posthumous book presents the evidence which finally caused him to decide that with ants the whole matter of caste formation is primarily controlled by heredity. Oddly enough this may well have been a mistake.

The method of caste formation is a question that will undoubtedly occupy students of ants for years to come. The evidence is not all in, and the indications we have that ant castes are determined by heredity make all the more interesting the instances in three separate kinds of social insects of the apparent evolution of group control of castes after the hatching of the egg. To this hasty sketch of the operation of group determination of caste in wasps and bees may be added that of termites.

The bees and their allies belong to one of the most specialized of insect orders, so that they are assigned a high position in the evolutionary tree of that class of animals. The termites, miscalled white ants, belong to a relatively unspecialized insect order related to the cockroaches, and stand low in the evolutionary scale among the insects. They have, however, reached a high state of social development.

Unlike bees, ants, and wasps, the termite colony, as we have said, is at all times composed of males and females in approximately equal numbers. There are male and female reproductives, of which three different kinds are known; these are the so-called first form, which have wings for a time and engage in a nuptial flight, the second form, which have wing buds, and the third form, which are wingless; and there are the sterile workers and soldiers, in which both sexes are also represented equally. The colony is usually composed of reproductives of some one sort and the two sterile castes (Plate V).

The controversy as to whether caste formation is a result of heredity or of the social environment has been as intense with students of termites as among students of ants. The trend of present information tends to support the theory of control by the environment.[31, 85] A certain California termite called *Zootermopsis* has reproductives and soldiers in its colonies, but no workers in the accepted sense of the term. Their place is taken by the younger nymphs, all of which have the possibility of developing into one of the reproductive grades or into soldiers. When Dr. Gordon Castle[31] set up experimental colonies of nymphs alone, he obtained in due time one or more pairs of reproductives. If the small experimental colony lacked a fertile male, one of the nymphs developed into that; if a fertile female was lacking and a male was placed in the colony, a nymph developed into a fertile female. If the nymphs in a colony that lacked both males and females were fed on filter-paper containing an extract of fertile females made with alcohol or ether, the males appeared at the usual time, but the females were delayed by twelve or sixteen days on the average.

Ordinarily in *Zootermopsis* only one soldier appears in the first year of the life of the colony. By removing the soldier as soon as it appeared in the experimental colony it was possible to get as many as six soldiers within the time that would ordinarily have yielded only one.

PLATE V. *Winged reproductive caste, soldiers, and workers of a termite from British Guiana. This is one of the largest species of termites and is shown life-size. A, winged reproductives; B, soldiers; C, workers. (Photograph by William Beebe.)*

In explanation of these and other similar data Dr. Castle expresses his opinion that at the time of hatching all nymphs possess three sets of possibilities to the same degree; namely, they may become sexually mature though wingless, or sexually mature and winged, or they may become soldiers. At some stage these chances are narrowed to two possibilities: the nymph may become sexually mature or it may develop into a sterile soldier. Since the reproductive possibility is present in all nymphs and since its expression is inhibited by a substance produced by a functional reproductive that affects the nymphs, the absence of functional reproductives would allow this potential power to express itself. Just what determines that one of the first small lot of eggs will become a soldier is not known, but it can easily be seen that when one soldier has started to develop it too may give off an inhibiting influence and so prevent other nymphs from becoming soldiers. In the normal course of events a second soldier appears only when the colony has become sufficiently numerous so that the soldier-inhibiting substance is spread among so many that the effect on any one nymph is weakened; and something of the same effect of numbers may explain why, in a large colony, many nymphs develop at times into sexually mature and winged forms.

There seems to be a relation to the more generalized situation noted earlier. When many animals are exposed together to a given amount of alcohol or some other toxic material, no one of the many may receive any overdose, as will certainly happen when one or a few individuals meet the full effect of the poison. This type of relatively simple mass effect, first discovered in experiments on group physiology among animals that at the most are only partly socialized, apparently turns out to be an important mechanism in regulating caste formation among these highly social termites; and some similar mechanism may control the activity of worker bees in producing new queens. It is true that the control of caste production is probably not the simplest form of physiological mass action, for the insects may from time to time become less sensitive to such inhibition. At these times, many of the nymphs may develop into the winged reproductives that swarm forth in the nuptial flight.

As many know, most termites eat wood, which, paradoxically enough, the more primitive ones are unable to digest although they do obtain their nourishment from it. The answer to this riddle is that these termites harbor in their alimentary canals several species of flagellate protozoans, which can and do change the wood into substances that both termites and these flagellates find highly nutritious.

From many structural relationships we know that termites are close relatives of cockroaches, and studies by Dr. L.R. Cleveland[38] have shown how the termite societies may have arisen from the much less social cockroaches. Here we have an example of one of the many possible connections between highly developed social life and the less social state illustrated by the mass physiology characteristic of animal aggregations.

Cryptocercus is a wood-eating cockroach found in decaying wood of the forests of the Appalachian mountains from Pennsylvania to Georgia, and along the coastal mountains in the northwestern part of the United States. Like their relatives the termites, these cockroaches feed on wood, and also like the termites they harbor wood-digesting protozoans in their alimentary tract. These wood roaches and many termites cannot live long if deprived of their associated protozoa, as can be done by appropriate treatment in the laboratory.

The young of both cockroaches and termites hatch out without these essential protozoans. The termites obtain theirs by swallowing a drop of liquid that has just emerged from the anal opening of another termite; the cockroaches get their protozoans by eating the pellets passed from the alimentary tract of molting individuals. Once a cockroach obtains a good supply it renews itself. One such cockroach or a pair can emigrate to a new log and live there for a lifetime. Since, however, adult cockroaches do not molt, the young of such an isolated pair, when hatched, could not receive the so-necessary intestinal protozoa, and hence a pair, if isolated, could not found a new colony. Actually the eggs hatch at just about the time of the annual molting season when the young growing roaches cast their outer covering and a part of the lining of their alimentary tract. At this time the newly hatched young can obtain protozoa readily and thereafter they retain them. The habit of living together is necessary in order that the newly hatched nymphs may acquire the useful protozoans from the growing, molting young.

The social situation is still more necessary for many termites. With them all the intestinal protozoans are lost with each molt, and each time that happens each newly molted individual must obtain some of the protozoans from another member of the colony or it will starve. The newly hatched termites often obtain protozoa before they are twenty-four hours old, and an artificially defaunated termite, if allowed to associate with his normal fellows, is reinfected within a few days. With such termites, colony life is an absolute essential, and only the winged males and females, the first-form reproductives already infected with protozoans before taking the nuptial flight, can even start a colony without the presence of others to carry the needed cultures of protozoans.

Many cockroaches that neither eat wood nor harbor wood-digesting protozoans reproduce so rapidly that given good hiding places and plenty of food they aggregate in large numbers, as many housewives know. These cockroach aggregations, which appear to be formed as a result of automatic reactions to the environment, accompanied by toleration for the presence of others, permitted the wood-roach *Cryptocercus* to develop the habit of passing protozoa from one individual to another, and so began the long evolution that has resulted in the highly adapted, wood-eating roaches found today.

The same basic adaptation allowed their relatives, the termites, to start on the much longer road they have traveled to reach their present state of highly developed social life.

We cannot outline the steps taken very closely, but it would seem that, in this cockroach-termite stock, aggregations allowed aspects of mass physiology to develop that in turn permitted a closely knit and varied social evolution. This is about as near as we have yet been able to come to charting a direct and obvious truly social development from a slightly social or subsocial animal aggregation.

Among grasshoppers, crowding can produce obvious structural changes. Certain species of grasshoppers found in semi-arid regions, such as those of South Africa, have two phases[4] that are quite distinct from each other. The phases are sufficiently different so that in the past they have been described as being different species. There is at present much evidence indicating that we can turn the phase *solitaria* into phase *gregaria* by crowding the young nymphs into dense masses. The opposite transformation may take place when the nymphs of phase *gregaria* are reared under uncrowded conditions. The differences between the two extend into color, form, and size.

Similarly, plant-lice, which are also called aphids, exist in winged and wingless forms that tend somewhat to alternate. When the wingless aphids have approximately exhausted the juices from one food plant, the next generation appears with wings; in flying about, some of them will usually find a new and suitable food plant on which they can settle and carry on. With some species one of the most effective ways of keeping wings from developing is to isolate the individual aphids and, conversely, one of the best recipes for obtaining winged forms is to allow them to become crowded.[114]

These distinctly different types of grasshoppers and aphids roughly suggest the structural differences between the castes of social insects, just as comparison was suggested between the structural differences of caste and sex. The resemblance is so close that the line cannot be drawn between its manifestations in social and infrasocial animals. Not only that, but the mechanisms by which the castes are produced appear in many instances to be like those that may occur when animals are aggregated together, even though the aggregations are below the level usually regarded as marking the lower limit of truly social life.

And since no one has yet demonstrated the existence of truly asocial animals it is impossible to define the lower limits of subsocial living. All that can be found is a gradual development of social attributes, suggesting, as has been emphasized throughout this book, a substratum of social tendencies that extends throughout the entire animal kingdom. From this substratum social life rises by the operation of different mechanisms and with various forms of expression until it reaches its present climax in insects and in vertebrates including man. Always it is based on phases of mass physiology and social biology that taken alone and in simplest development seem to be social by implication only.

Some Human Implications.
Chapter X

While we have been engaged in trying to assay the relative importance of the principle of cooperation among animals, we have given most of our time and attention to its manifestation among animals considered to be asocial or only partly social. In such animals it is an unconscious kind of mutualism, but its roots are deep and well established and its expression grows to be so spontaneous and normal that we are likely to overlook or forget it in the more striking exhibition of social cooperation among higher animals. Conscious cooperation at the level of psychosocial facilitation is so comparatively new in an animal world many millions of years old that we may underrate its strength and importance if we are not reminded of its foundations in simple physiology and primitive instinct.

When we attempt to apply to human behavior the same methods of analysis that we have used throughout toward other animal groups, we reach most interesting results when we select some phase of reactions of men in which integration has not developed much beyond that found in some of the semi- or quasi-social animal aggregations that we have been considering in the lower animals.

Among the possible aspects of human behavior that meet this requirement and lend themselves to biological analyses is the whole set of activities that center about the relations between nations. Even the most optimistic humanist will not maintain that these are at present, or ever have been, on as high a social plane as that which characterizes many of the personal interactions of mankind, or those of the smaller social groupings of men.

The most casual reading of recent events is convincing evidence that the modern international system is based on war. This final resort to violence has been regarded by many thoughtful people as inevitable, man being what he is, that is, the product by natural selection of the results produced by the struggle for existence; for the ordinary thoughtful person is not aware that the tendency toward a struggle for existence is balanced and opposed by the strong influence of the cooperative urge. Because of this common attitude toward war, and because of its fundamental importance to our species, I propose to cut through the shifting tangle of international policies down to the basic biological significance that it holds for us.

In doing so I must recognize these two fundamental principles, the struggle for existence and the necessity for cooperation, both of which, consciously or unconsciously, penetrate all nature; and I shall say now that we may find that these two principles are not always in direct opposition to each other; there is evidence that these basic forces have acted together to shape the course of evolution, even the evolution of social relations among men and nations of men.

If, in the past, we have not had facts on which to base rational conclusions about national problems, it cannot be said that we have not had powerful emotions to drive us into one attitude or the other. It is very difficult to keep an objective, unemotional attitude toward the complex subject of the biology of war. You and I may not agree in our placing of the emphasis, but I trust that when we disagree it will be on a healthy intellectual level.

It is clear that we are entering a tricky field where, to a greater extent than usual, the evidence is not all in, and one in which much that we think we know is contradictory. No one can bring this problem into the laboratory for careful testing. We must do the best we can with information that is more incomplete and faulty than that on which biological discussions are normally based. The human importance of the subject justifies the risk. The present discussion will center about three main points:

1. To what extent do the underlying biological relationships tend to bring about war?

2. Is war biologically justified by the results produced?

3. Can the basic principles of struggle and of cooperation work together in the international relations of men?

Many men are aggressive animals. The similarities between human social hierarchies and those of chickens and other animals emphasize similarities of the drive toward dominance in the species concerned. Our immediate question is: Does this human aggressiveness mean that men have an inherent, instinctive drive toward war? The ideal way to attack this problem would be to rear sizable groups of people free from contact with outside influence or social tradition and see whether under these conditions they would instinctively engage in group combats in order to forward or defend group ambitions.

Such objective procedure is out of the question, but an interesting subjective inquiry has been made. In 1935, American psychologists took a poll among themselves on the question of whether they believed that the tendency toward making war is an instinctive drive in man. Of those answering, well over 90 per cent said that there is no proof that war is an innate behavior pattern.[125] Less than 10 per cent thought that war represents an instinctive reaction. I did not personally see this questionnaire, but I am credibly informed that the question was stated fairly and did not suggest the type of answer expected.

This is a rather unexpected unanimity, and may be accounted for to some degree by the existence of one modern school of psychologists who doubt the possibility of instinctive action, particularly among men. I do not think they represent a large proportion of American psychologists, but there may have been enough of them to have lifted the percentage high.

The opinion of the psychologist is supported by the independent judgment of one of the leading recent students of anthropology, Professor Malinowski, who said in his Harvard tercentennial lecture:[88] "All the wrangles as to the innate pacifism or aggressiveness of primitive man are based on the use of words without definition. To label all brawling, squabbling, dealing of black eye or broken jaw, *war*, as is frequently done, simply leads to confusion. War can be defined as the use of organized force between two politically independent units, in the pursuit of tribal policy. War in this sense enters fairly late into the development of human societies." [141]

It is not impossible to break down and remake instinctive behavior, as the change in marriage customs since the days of the cave men shows us. Nevertheless, it is much easier to change learned behavior patterns, one of which these experts believe war to be.

We must still take account of individual aggressiveness, and the fact that man appears to be relatively easy to lead into mass combat. Even if war-making is not instinctive, if it is a learned pattern of social behavior, there is evidence that it has existed for some fifty centuries, and it would probably require at least a few centuries of intelligent and fairly concerted effort by those who do not believe in its utility for man to unlearn the habit.

There is a second important set of biological processes that at first sight appear to work inevitably toward the production of war. These center about the question of overpopulation: that is to say, about the relation human numbers bear to habitable land areas. This is the next primary problem which we must consider.

Over the world there is a limited range of habitable land; and thus far we have no intimation of any practical method of emigration to neighboring and perhaps less occupied planets. And there is a rapidly expanding human population, which is even now becoming uncomfortably dense in the crowded nations. It is often said that this is a fundamental cause of tension that makes wars inevitable, as hard-pressed dense populations seek food in more amply-provided areas.

The desirable biological results of wars so induced have been, and still are, supposed to be two:

1. The dense populations are thinned to the bearable point as a result of the fighting, or

2. Superior nations, or races, are victors. They expand at the expense of the defeated inferior group and so occupy more of the limited space that is available for men.

Let us test these theories against the known facts, using for that purpose the somewhat seasoned figures collected and analyzed in the years between the two world wars. Roughly speaking, there are about fifty-two million square miles of land surface on the earth.[106] This includes the habitable plains of the temperate regions; it also includes the relatively uninhabited deserts, tropical jungles, and mountains. Approximately one fourth of these fifty-two million square miles is desert or semi-desert and can support only a sparse population of men. This leaves roughly forty million square miles of non-arid land theoretically open to human habitation.

On this land there were living in 1935, according to a revision of the estimated world population made by the late Professor Raymond Pearl, something over two thousand millions of people. This is almost exactly forty people per square mile of the whole earth's surface, or about fifty people per square mile, if the arid and semi-arid land is excluded. We can better visualize the meaning of these figures when we know that they approximate the average population density for the United States; forty per square mile for the whole land area, and fifty per square mile on land with fair rainfall.

An estimate of human population of three hundred years ago, tentatively advanced by Professor Pearl, is that in 1630 there were probably about 445 million people on the whole earth, or about eight per square mile of total land surface.[106] Dr. Pearl thought that this was probably the largest human population the earth had supported up to that time. Then came the opening of the Americas for settlement, the beginnings of modern use of transport and manufacturing processes, and the scattering of information by modern methods. The result has been that in the last three centuries the population of the world has increased almost fivefold, from eight to forty per square mile, largely because food and shelter and mechanical energy were made available for five times as many people and because the development of modern science made the world safer for them.

In these three hundred years the world population has doubled, on the average, approximately every sixty-four years. Today mankind is increasing in numbers at such a rate that *if the increase should continue* as it was going in

1935 we could expect another doubling of the number of people in the world in the succeeding seventy years, and we should have about eighty people per square mile in the year 2005.

What then? Will not the coming generations at some time be obliged to fight for their place in the sun?

This prospect is somewhat altered, however, by the fact that many students of population trends believe that the rate of human increase is slowing down. Thus, in the case of the United States, a doubled population by the year 2005 over that in 1940 would call for some 262,800,000 people living within our boundaries. In contrast, Dr. Baker,[24] economist in the Department of Agriculture, estimated in 1937 that the maximum population to be expected, if *conditions remain unchanged*, would be less than 150,000,000, a population density that has already been reached!

Experts differ concerning expected population size, but not about the prospective slowing of the rate of population increase. Professor Pearl and his associates after the 1910 census estimated a total of more than 197,000,000 people in the United States in the year 2100, but following the 1940 census they revised this figure downward to about 184,000,000,[11] or slightly fewer than seventy people per square mile of the total land area. I know of no expert who expects our American population to double itself again unless there is a radical increase in available energy or in other aspects of our living conditions.

For the world as a whole, Dr. Pearl estimated in 1936 that, *if present trends continue,* as they may not, the world population will reach a maximum of about 2,650 millions by the year 2100.[106] This is a density of about fifty persons per square mile of land surface on the globe, counting good and bad land alike.

I must dissociate myself from any responsibility for

these and similar estimates. I fully realize, as do their authors, the pitfalls inherent in such predictions. Human trends being what they are and have been in the last three hundred years, this is as good an approximation as can be made at present, and with all its faults it is worth considering.

The important aspect to me is that we do not have reason to expect in the United States or in the world a continuation of the unprecedented rate of increase of the last three centuries, or even a continuation of the present rate of increase. Unless population experts are all at fault, the rate of reproduction among human animals is slowing down, just as the rate of increase in nonhuman populations slows down as laboratory containers approach an overcrowded condition. In fact, few animal populations approach the limits of their food supply in nature.

The reasons for this are not clear, though they appear to be connected with the ease of securing available energy, food, and shelter. As men approach the bearable limits of these necessities of life there occurs an increase in birth control. This is shown in Italy, where, according to figures given in *The Statesman's Year Book*,[126] despite continued propaganda for a higher birth rate the actual number of births fell over 12 per cent from 1922 to 1936 (Figure 47). Thanks to a similar decline in death rate the significant percentage of births which are canceled by deaths has remained fairly steady. In England, where there has been no great effort to encourage population increase, the deaths in 1922 were 62 per cent of the births; in 1935 they were 81 per cent. Perhaps the success of Italian efforts is to be measured by this comparison with England rather than by the fact that under propagandist pressure the birth rate actually decreased. In Germany, the post-Nazi regime has not been in existence long enough to establish a trend. The graph (Figure 48)

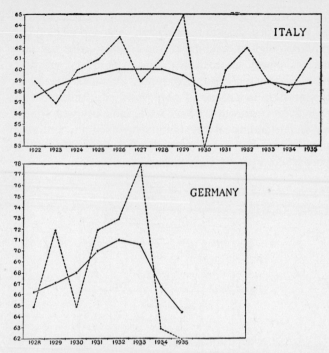

FIGURE 47. *The percentage of births that were canceled by deaths for eight given years in Italy and Germany. The higher the trend line, the slower the population is growing, and vice versa. The broken line connects the observed points; the solid line shows the mathematically smoothed trend line. (Data from* The Statesman's Year Book.)

shows that beginning in 1934 there has been a dramatic decrease in the percentage of births canceled by deaths; actually there was a decided increase in births. Analyses in the *American Journal of Sociology* indicate that the increase in the birth rate just after Hitler came to power may have been the result of a campaign against abortion that in pre-Nazi times terminated over one third of the pregnancies.[66, 137] One can deduce from general biological experience, despite the German data, and lacking cultural or religious teaching to the contrary, that the population almost automatically adjusts numbers within its physical and biological limitations. Doubtless eventually this mysterious process of population adjustment will be analyzed. At present we have made some progress toward an understanding of the factors involved in nonhuman populations, but have little objective knowledge to report where men are concerned.

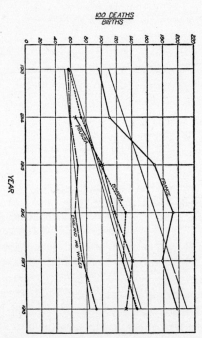

FIGURE 48. *The percentage that deaths were of births steadily increased during the First World War in France (non-invaded departments), Prussia, Bavaria, England, and Wales. (From Pearl.)*

It is of course possible to increase the present food supply of the world enormously. It has been estimated that if our present biological knowledge were consistently applied we could raise food enough to supply at least ten times the present world population, instead of the 25 per cent increase to which we are looking forward by the year 2100. Presumably by that time we shall have learned much more than we now know about intensive methods of food production.

Let us take one simple instance only. In the United States we are substituting gasoline-driven farm machinery for horse power in agricultural work.　The land required to produce feed for one horse will equally well provide food for a man. Baker, the agricultural economist cited earlier, estimates that the land released annually by this change in farm technique can be turned to growing human food almost as fast as our population is increasing.

The question seems rather one of adequate food distribution than of shortage of food. Under conditions that we can visualize at present there seems little likelihood of a real food shortage for the world as a whole.

If, however, these conclusions prove to be completely wrong, and the world population is now or will become too high by biological standards, there is still the question of whether war is a sound and sufficient means of controlling population growth. The theory that war is an efficient means of stopping the increase of mankind is so contrary to fact that I allow myself to say No in the first place and present the evidence later.

The usual effect of a war upon the civilian population is to depress the birth rate and raise the death rate on both sides of the line, whether in the winning or the losing nation. Figure 44, taken from a study by Pearl on population trends during the First World War, gives these data for the unoccupied parts of France, for Bavaria, and for England, from 1913 to 1918.[103]

In 1913 deaths and births in these parts of France were almost equal; in 1918 there were approximately two deaths for each birth. In Bavaria in 1913, there were five births for every three deaths; in 1918 there were three births for every four deaths. The trend lines in Figure 48 for these two countries run almost parallel, though France was invaded and losing in much of the fighting while Bavaria was free from foreign troops and part of a winning nation until near the end. As usual, analysis of such a situation is not simple. Bavaria, although enjoying the psychological advantage of belonging apparently to the winning side, suffered the physiological disadvantage of an increasingly severe food shortage, while France averaged an adequate food ration. In England during the same time, where there was neither invasion nor starvation, there was the same tendency toward increase of deaths in proportion to births, though less marked. These statistics, of course, do not take into account the almost unprecedented death rates in the fighting lines.

Temporarily the population growth was checked, but almost immediately following the close of the war the ratio of births to deaths resumed the prewar trend lines. Pearl, writing in 1921,[104] summed up his study in these words: "Those persons who see in war and pestilence any absolute solution of the world problem of population . . . are optimists indeed. As a matter of fact, all history tells us, and recent history fairly shouts in its emphasis, that such events make the merest ephemeral flicker in the steady onward march of population growth."

Fifteen years later, in 1936,[105] Pearl wrote, alluding particularly to the effects of wars of conquest by one nation to acquire the territory of another: "The world problem of population and area, however, remains unaltered in theory, though practically it will have been made worse because of the extravagantly wasteful destruction of real wealth that war always causes. This is the problem that is really serious—how can forty persons be maintained for every square mile of land surface of the globe— good, bad and indifferent land together? War cannot enlarge the land surface that must support mankind; it has never diminished the total number of people who want to live on it except by a tiny fraction for quite a brief period. There is no way out of the dilemma by the pathway of war."

It is a comparatively new idea that population can be controlled at all except by famine, pestilence, and war, which have been regarded as acts of God. Acts of God or not, we can no longer tolerate famine or pestilence if we have the power to prevent them; and lacking such power we intend to get it as soon as it is humanly possible. Among dispassionate, expert students, war has similarly lost caste as a means of population control, though the man in the street has not yet learned this.

Instead of the dubious check these agencies furnished there is a steady turning to birth control, even in the countries where it is most surprising to find this, although all too slowly in such populations as those of India, China, and Japan.

There is significance not only in the average density of people per square mile of the earth's surface but also in the population density of the most crowded nations. The degree of crowding in 1936 in certain countries with whose problems we are familiar is shown in the following list. The figures given are slightly rounded statements of the average population density per square mile of land territory. The most densely populated countries of the world are listed here in order:[105]

COUNTRY	PEOPLE PER SQUARE MILE
1. Belgium	700
2. England and Wales	680
3. Netherlands	660
4. Japan	450
5. Germany	360
6. Italy	360
7. China (proper)	300
8. Czechoslovakia	270

For many purposes it is hardly fair to compare the relatively small countries such as Belgium and the Netherlands with others such as Japan or Italy, which are larger but contain a high percentage of waste land. For our purposes, however, the list as it stands is fair enough; such data represent the facts we have to face.

At present about two and a half acres are required to supply food to one person, if the soil is fair to good and the husbandry is good according to present standards. This means that under modern conditions of agriculture the upper limit of a relatively self-contained population is about 250 people per square mile. It will be seen that Belgium, with its 700 per square mile, almost triples this upper limit, and that England and Wales and the Netherlands more than double it. Such high population densities can be supported by trade conducted with other countries on a large scale. They could also, as we have seen earlier, be supported by improved methods of agriculture. An Italian expert on populations said in my hearing some years ago that population pressure is not a direct cause for war, but can be used by a clever leader to range a nation behind aggressive policies that lead to war. In the short run such procedure is easier than to educate people to apply the available knowledge that would allow

Italy, for example, to feed her present population, and more, from the products of her own soil.

It is time now to turn to the second of the questions concerning the biological background of war. In the light of the preceding discussion we can restate this question as follows: Although underlying biological relationships do not necessarily lead to war, is not war biologically justified by the results produced?

If war does benefit the race in distinct and unique ways, then the biologist must favor a system of society that will bring about the proper kind and the correct number of wars to produce the best racial selection. If war, on the other hand, tends toward human deterioration, then the biologist must oppose a system of international relations based on war. Again it is a question of evidence.

The matter of *individual* biological selection is one that is fairly obvious even to the layman; and his conclusion that the direct results of war are harmful biologically has been well supported by scientists whose interest in the subject is more inclusive than their natural sympathy for the young men of their acquaintance who have incurred wounds or have been gassed or have suffered severely from some of the typical wartime epidemic diseases. The work of David Starr Jordan before 1914 is classic; [78, 79, 82] but the evidence furnished by the First World War is more important to us. American experience at that time is best set forth in the slender book by Professor Harrison Hunt,[75] who studied the records of the American army, using modern statistical methods.

Professor Hunt was left with no doubt that war selects the best of our young men for exposure to wartime hazards. We have space for one bit of evidence. Hunt found that, for the drafted American army in that war, 83 per cent of the mentally defective were rejected; those of normal mentality and the 17 per cent who were only slightly subnormal were held for service. A good geneticist interested in improving the human stock would have reversed the procedure, sending the mentally deficient out into wartime risks and keeping the others at home to continue the race. But this is so contrary to fact in all the standards by which armies are selected that it seems faintly ridiculous in the telling. Personal selection, so far as it exists in modern warfare, selects the individual to be killed or wounded because he is physically or mentally superior to those who are left at home.[72]

The ill effects of this selection among the young men are evident in a nation in which war losses have been heavy, but they are less drastic for people as a whole than they might be if it were not for various mitigating factors. To date only half the race has suffered in so-called civilized warfare, since even in the Second World War, women have been exempt from actual combat. Also, many young men return who, though wounded and perhaps otherwise handicapped, are still physically capable of passing on their germ plasm to succeeding generations. And even in populations badly shattered by war most of these genetic ill effects could be obviated if monogamy were less of an ingrained human practice.

The effects of severe wartime epidemics, which are usually the cause of more deaths than the actual fighting, are subject to the same comments; but with these epidemics the civilian population is also directly affected, as was the case with the influenza pandemic that swept the world in 1918 and carried off in a day more civilians than did many spectacular air raids combined.

General epidemics tend to fall most heavily on the old and the young; biologically we are most interested in the fate of children and young people. Disease and undernourishment drastically reduced the younger population in places well away from the fighting lines in the First World War. Homer Folks,[53] United States Red Cross commissioner, testified in 1920 that in some sections of Italy 60 per cent of the children failed to survive wartime conditions. The children of Germany and of Poland also suffered greatly.

If he could know that such severe exposure eliminated the relatively weaker specimens and left a stronger, hardier race, the biologist could reconcile himself to the death of these children, though emotionally he might rebel.

But this rationalization is impossible. Study of the after-effects of epidemics upon children[50] does not show a group of sturdy survivors, with all the weaklings eliminated. Rather, the later history of these children shows that they have a lower resistance to the next severe disease that strikes them. Apparently many such children, though surviving, are weakened for some years thereafter.

Similarly, the children back of the battle lines include many whose experience left a mark, and who recover only slowly from its injurious effects. They were not a selected lot, and their own generation suffered. Fortunately all our evidence indicates that those who survived are able to pass on their inherited qualities unimpaired to their children; but many are unable to provide for their families the physical care and conditions for living that make for the fullest development of inherited potentialities.

Perhaps a sane and cautious quotation from Professor S. J. Holmes will be a fitting summary for this section. In 1921, Holmes wrote:[71] "On the whole it is quite probable, I believe, that the effect of military selection is harmful. . . . It is a matter of serious doubt whether the beneficial factors come near outweighing the adverse selection of battles."

What are some of the beneficial effects which this

statement suggests may exist? One of them is that war is necessary to maintain racial vigor. This is a matter on which statistics are not available, and on which personal opinion must play as reasonable a part as it can.

To me it seems a misreading of history that leads to the justification of war as a means of keeping up the vigor of the race. I should say, rather, that wars have frequently revealed the loss of racial or national vigor among a people made soft by easy living, which in turn had been made possible, at least at times, by a long series of successful wars of conquest.

Anyone who attempts to maintain the thesis that wars do keep racial stocks vigorous—and there are biologists who believe this—is troubled by China. This much-discussed and frequently invaded land was populated by the forerunners of the present Chinese during the days when Egypt, Assyria, Babylon, Greece, and Persia were fighting the wars recorded in our general histories. Those war-like peoples have lost their racial vigor, but the Chinese, who have been relatively peaceful, have retained it. This stumbling block cannot be removed by denying racial vigor to the Chinese; they have, in the past, absorbed too many temporary conquerors, and have occupied and are occupying too much of the earth's territory, to be dismissed as a racially decadent people. There are anthropologists who reckon them biologically the most advanced people living today.

There is another allied but somewhat different theory regarding the human benefits conferred by war which holds that, even though in direct personal selection the war system is dysgenic, it does tend to select the fittest races and nations for survival. This theory is usually applied to European history, where in the long struggle of advanced European nations against backward, poorly equipped natives of America, Asia, Australia, and Africa,

victory has eventually rested with the Europeans. Whatever the intrinsic human merits of the case, a question
on which Hindus may disagree with Englishmen, there
can be no doubt that such conflicts have been won by
the nation possessing the more modern social organization
and the better gadgets with which to fight; and the
winning nation has not hesitated to levy on the weaker
one for whatever possessions and services it could utilize
for its own advantage.

When, however, two European nations fight each other,
as, for example, France and Germany, who can maintain
that either is biologically superior? Let us look at the
record since 1800. Germany was overrun by Napoleonic
armies and then helped to win at Waterloo and won
readily in 1870. France long held Germany in the Napoleonic era, lost at Leipzig, Waterloo, and Sedan, was
heavily invaded in 1914 and in 1940, and yet emerged on
the side of the victors both in 1918 and in 1945. Such results prove nothing to the biologist engaged in studying natural selection among peoples. Or, to come closer
home, does the fact that the Confederacy lost the war
between the States prove that the white people of the
South are racially inferior to those of the North?

Actually, of course, we are not fighting racial wars
at present. What race won the world wars, or, for that
matter, lost them? Modern warfare among so-called civilized powers probably does result in victory for superior
wealth, better organization, shrewder propaganda, and
other social achievements, but we have little good evidence to link these social attributes with racial stock.

Let us allow Popenoe and Johnson,[109] recognized students of eugenics, to summarize this whole inquiry into
the biological justification of war. Writing in 1918, when
the subject was near the top of men's minds, they said:
"When the quality of the combatants is so high com-

pared with the rest of the world as during the Great War,
no conceivable gains can offset the loss. It is probably
well within the facts to assume that the period of the late
war represents a decline in inherent human quality greater
than in any similar length of time in the previous history
of the world." A similar assertion holds for the second
Great War.

It seems to me that such evidence and reasoning as I
have presented indicates pretty clearly that the present
system of international relations is biologically unsound.
Attempts that have been made in the past to lend biological respectability to the existing system by regarding
it as an expression of an inevitable struggle for existence
have overlooked not only its defects as a selecting agent
but, more serious, have often not even been conscious of
the existence of another fundamental biological principle,
that of cooperation. Is it possible to envisage a system of
international relations that will be fairly based on both
these aspects of biology?

One of the first questions to be examined is that of the
size of the cooperating unit practicable in such a system.
It is possible to make a case for the present human social
divisions, whereby nations of various size cooperate within their own boundaries though competing with each
other for various types of supremacy. Within each of these
nations are graded series of groupings in great variety,
which also cooperate within and compete across their
tangible or intangible boundaries. Here immediately we
come across an important qualitative difference in the
competition. Within each nation this intergroup struggle
is normally carried on by approximately peaceful and
orderly means. By contrast it is accepted that the competition across national limits, usually peaceful and orderly, may at any time break down into the socially
backward phenomenon called war; and even in periods of

peace and social progress much of the average nation's energy, wealth and forethought is diverted to preparing for the next war.

Peaceful intergroup competition within a nation has come to rest, in the first place, on habit, preference, and a realization that only temporarily is an advantage gained by violence; and, in the second place, on a government, often set up by mutual consent of the competing groups, which is strong enough to block or stop cruder appeals to force, and which is expected to do so.

The suggestion has been urgently repeated since the time of Sully,[69] the great minister of Henry of Navarre and France, that there should be a similar international organization. Theoretically there is almost everything to be said for this proposal. Such an international organization might be set up much as the federal government of our country was planned, to supervise the functioning of the different states. This system calls for representative government, a relatively unbiased court of final judicial appeal, and certain potential police power, which in our American experience has been used but rarely on a national scale.

The League of Nations, even in its most hopeful days, did not show more than remote possibilities of equaling on a world scale what the British Empire did fairly adequately, through the nineteenth century and longer, for more than one fourth of the earth's land area. The United Nations, or any future international body that will undertake to apply the balanced principles of struggle and cooperation on a global basis, must, among its other qualifications, avoid certain outstanding mistakes of the defunct League.

An international organization cannot be really cooperative if it is basically a league of victor nations formed to administer a punitive peace treaty, for this is hardly a step in advance of the time-honored national alliances for defense and offense, which are cooperative only to be destructive. It must not be dominated in any department by the representatives of any one nation, not even when that nation is as intelligently, and shall I say selfishly, benevolent as England and its dominions were at their very good best. It must be so organized as to secure and hold adherence from the great majority of nations. As a step toward this end, the biologist's international system must be a dynamic organization capable of and designed to effect changes rather than set up to preserve any given status quo, regardless of how favorable for the predominant powers.

Biology teaches the inevitability of change, if it teaches anything. We must have some device in our system which will allow for needed changes, some means of making those compromises at which the English and the French are so proficient in their internal affairs. In international as in legal circles, we must have some peaceful means of declaring a defunct nation to be in fact bankrupt or unable to manage its own business, and to distribute its assets among the proper creditors.

When such a system is installed there will need to be not only the means for international consultation, and a hearing for the troubles of the world; there will also be a necessity for courts of international justice. One of these may well be patterned on the Supreme Court of this country; another might be a development of the international court of arbitration that has been located for many years at The Hague.

At this point we come to a serious divergence of opinion. Should these courts be supported by police power? As a realistic biologist it seems to me that international police force will probably be a necessity in those cases in which a nation or a section of a nation attempts to raise itself in the peck order of governments by direct action rather than waiting for the results of the more just but slower pressure of world opinion. Much of the police activity should be limited to such duties as are now exercised by our federal marshals, but in my judgment there would need to be the possibility of the use of even stronger police pressure.

It is certain that if an international organization is to succeed, police power must be used very rarely. The attempts of the British government to coerce the American colonies or the Irish people are conspicuous as a demonstration of the frequent failure of massed force to compose complex human maladjustments. It is noteworthy that such enforcement has not been used in the long and successful operation of our own Supreme Court.

Practically, it is possible that nations will join in an international enterprise that is limited to consultation and judicial review of all disputes long before they will relinquish any other phase of their jealously guarded sovereignty to such an international organization. We may even be able to work out a method of international cooperation based entirely on patience, wisdom, and justice, though in the light of past experience this seems at present unlikely.

Such a world organization will never be perfect. Man is not. Neither is the government of Chicago, of Illinois, of our United States. And yet who would not prefer to live in Chicago, even back in the gangster era of the nineteen-twenties, rather than in the period of greater individual freedom for privileged people that London or Paris of the Middle Ages afforded?

A thoughtful and sincere biologist may object that the world is too large an area for a successful cooperative unit; that we need units intermediate in size to allow for human evolution those advantages that are found in populations intermediate in size. To such an objection one must reply that, as to the latter point, the maintenance of smaller cooperative and competing units within the larger one is part of the scheme as sketched. And to the first point, that of the great size of the earth, it need only be mentioned that, thanks to recent improvements in transportation facilities, New York is in point of time as near the Orient as it was to Los Angeles in 1885; and there are few places on the globe as remote from Washington as was San Francisco before the Union Pacific Railway was built. In transportation and communication, and in community of essential human interests, the world is ripe for a workable international organization.

From the standpoint of pure biology, disregarding considerations that may seem to smack of the social sciences, the mortal enemies of man are not his fellows of another continent or race; they are the aspects of the physical world that limit or challenge his control, the disease germs that attack him and his domesticated plants and animals, and the insects that carry many of these germs as well as working notable direct injury. To the biologist this is not even the age of man, however great his superiority in size and intelligence; it is literally the age of insects.[5]

This is a fact which must have repeated emphasis. In the tropics there is only the narrow strip along the Panama Canal and similar small areas in which man has shown the ability to compete successfully with the insects; and the techniques of this competition are too expensive as yet to apply along the vast rich stretches of the Orinoco River, the Amazon, or the Congo; there, undoubtedly, the insects are in control. In countries such as India and southern Russia, mosquito-borne malaria is a plague that saps the energy of those enormous populations as it does today in the Caribbean region.

There are good biological precedents for such competition between different types of organisms as that between man and insects or between man and bacteria. In fact, with almost negligible exceptions, the only kind of mass slaughter for which there is precedent in animal biology is found in interspecific struggles. One species of animal may destroy another and individuals may kill other individuals, but *group* struggles to the death between members of the same species, such as occur in human warfare, can hardly be found among nonhuman animals.

These techniques by which we can successfully combat our enemies, the insects, and the disease-producing bacteria, protozoans, and viruses they transport are too expensive for the world today. They are too expensive because even the peaceful nations are using so much of their resources for buying and building armaments on an unprecedented scale, apparently to make one more experimental test of the fact that war is biologically indefensible.

In our struggles with our physical environment, with disease germs and insects, we have ample opportunity for the struggle for existence, and stimulus enough to apply to the limit the principle of cooperation.

Unconsciously or consciously, the innate urge toward cooperation appears even under circumstances where it would seem least likely to be fostered.

Even in the most seriously war-torn countries, as in Spain during its most recent revolution or in the various countries in which the battles of the Second World War were fought, when one was withdrawn from the actual scene of battle one found the common people engaged as best they could in their normal activities of providing food, clothing, and shelter for themselves and their families, with the ineradicable drive toward constructive cooperation that we have found evident throughout the animal kingdom. Such cooperative activity will reach through a family, from family to family, from city to city and even across frontiers.

These normal activities can be wiped out in a few minutes by the exaggerated expression of the struggle for existence that we call war, extended beyond all biological justification and become, as Malinowski has said, "nothing but an unmitigated disease of civilization."[88]

It is a disease of long standing that even under most favorable conditions we must not expect to see cured overnight; but the outlook is not without hope. There seems to be no inherent biological reason why man cannot learn to extend the principle of cooperation into the field of international relations to as great an extent as he has already done in his more personal affairs. In addition to the unconscious evolutionary forces that play on man as well as on other animals, he has to some extent the opportunity of consciously directing his own social evolution. Unlike ants or chickens or fishes, man is not bound over to form castes or peck orders or schools, or to wait for a reshuffling of hereditary genes before he can discontinue behavior that tends toward the destruction of his species.

The Peck Order and International Relations. Chapter XI

The need to focus knowledge from every available source on the problems of the world today is especially pressing in the light of existing international tensions. It is particularly important to examine underlying principles, against which such troublesome matters as the roles of China, Russia, and the United States in world affairs, may fall into perspective.

It may have come as a surprise to some that there is evidence from modern experimental studies in group biology that bears on international problems. The biological information at hand is incomplete and must be used with caution, but the urgency of the situation has seemed to me to necessitate breaking through the reticence of the research biologist to set forth, and even to summarize briefly, some of the human implications growing out of recent work with animal aggregations.

Intensive research in the last few decades has verified the long-known principle that overcrowding produces harmful results, in part because of intense competition among the crowded animals. These harmful effects are said to illustrate what may be called disoperation. In contrast, all through the animal kingdom—from amoeba to insects, or to man—animals show automatic unconscious proto-cooperation or even true cooperation. There is much evidence that the drift toward natural cooperation is somewhat stronger than the opposing tendency toward disoperation.

Recent experimental findings confirm the insight of earlier biologists—Espinas, Kropotkin, and Wheeler among them—that "living beings not only struggle and compete with one another for food, mates, and safety, but they also work together to ensure to one another these same indispensable conditions for development and survival." [132] This natural mutualism is one of the great principles of biology. It is as important as it is undramatic.

There is a contrasting principle, characteristic of the group biology of a few invertebrates and of most vertebrates, that is much more dramatic, though less fundamental. Groups of fishes, lizards, various birds, and many mammals including man are often organized into social hierarchies. The social order usually rests on threats or on the direct use of force. The hierarchy in a flock of hens, for example, grows out of the right of the more dominant hens to peck their social subordinates without being pecked in return. This right is won by fighting, unless each successive opponent submits without a combat. Thus the social hierarchy of hens is based on a "peck order." Typically, in a small flock of hens, A pecks B; both peck C; all three peck D; and so on to the omega hen, which is pecked by all others in the flock. High position in the peck order gives more ready access to food, space, mates, and to other things that are important to hens.

Small-scale human societies are more closely coordinated than are most groupings of other animals, but when nations form the individual units we find a laxity of integration somewhat similar to that shown in animal aggregations. There is also a strong tendency to form international peck orders, based on actual or potential power, that resemble, at least superficially, the hierarchies of hens and other vertebrates.

Much can be said for an established order of dominance and subordination, whether within groups of nonhuman animals or among nations. There is growing evidence that with hens, again as an example, well-organized flocks, in which each individual knows and is fairly well resigned to its particular social status, thrive better and produce more eggs than do similar flocks that are in a constant state of organizational turmoil. Similarly among nations, relative quiet exists when the international order of dominance is fairly firmly established and generally accepted. Evidence may be taken from parts of the long period when Rome dominated the western world of her day, or, more recently, when England and the growing British Empire dominated the post-Napoleonic scene.

Sooner or later, however, on the international stage as among our groups of mice, or fish, or hens, or other animals, a subordinate always seriously challenges the alpha individual or nation. Although the challenger may be beaten back, often many times, eventually alpha rank is taken over by a new despot, and the cycle starts again.

In so far as any international organization, formal or informal, is based primarily on a hierarchy of power, as are the peck orders of the chicken pens, the peace that follows its apparent acceptance will be relatively short and troubled. Permanent peace is not to be won by following the precedents established by the dominance orders of vertebrate animals.

From a political rather than a biological approach, the

federal government under which we live in the United States was set up in order "to form a more perfect union," not to establish Massachusetts, New York, Pennsylvania, or Virginia, or any combination of them, in alpha rank in the new republic. The continuing success of the United States as an integrated unit is closely related to the lack of emphasis upon possible peck orders among the states.

In contrast to the implications drawn from animal hierarchies based on aggression, the study of natural cooperation and its forerunners among nonhuman animals furnishes suggestions for lasting stability, at least in limited phases of international affairs. A few brief summarizing statements will illustrate some of the possibilities:

1. Many animals care for their young even at the risk of their own lives.

2. When food is plentiful, fighting among animals normally decreases, but it decidedly increases with food shortage.

3. Widely varied assortments of animals, predators and prey alike, often ameliorate their common environment to the advantage of all.

An international organization would have a real chance for immediate and lasting success if—except for large over-populated areas steadily showing too high a rate of increase—it devoted itself to the following activities:

1. To seeing that children throughout the world receive adequate care, sufficient food, and modern schooling.

2. To ensuring the production of plentiful food and its distribution.

3. To improving living and working conditions, especially among low income groups.

So far as they have been practiced to date, each of

these biologically sound, essentially cooperative activities has succeeded on a limited international scale; for example, the International Labor Office was perhaps the most successful phase of the Geneva League of Nations. An organization of the nations of the world devoted primarily to meeting these and similar human needs would be based on the great drive toward natural altruism that extends throughout the whole animal kingdom.

Given practice in cooperation between national governments, and our human additions to the tendency of all animals to cooperate, we may be able to work out an adequate control even for the prestige problems of the international peck order. It is encouraging to remember that, within most nations, we already manage complicated group and class peck orders with relatively little use of force. And, at the individual level, unlike the situation with many nonhuman animals, or with nations, fighting is no longer approved as a method for determining personal status in the varied social hierarchies of men and women. Rather, more or less cooperative practices prevail.[7]

Summary.

Chapter **X I I**

Interest in the social impact of science in general and of biology in particular has been growing steadily in the past few decades. There have come the terrific impact of the atom bomb and the dread potentialities of the proposed hydrogen bomb. But more dangerous than fission bombs and other achievements and prospects of physical scientists is the power that may arise from scientific mismanipulation of men's minds and devotions.

My own active concern with various phases of the subsocial and social life of nonhuman animals has revealed enough of the complexities of these simpler social systems to make me well aware of my limitations when confronted with the modern social problems of men. It is the drive of immediate necessity rather than a feeling of competence that impelled me to undertake the discussion of the biological foundations for some fundamental phases of the social behavior of men.

In our laboratory we are making two experimental approaches to the phenomenon of biological sociology, and each yields its very different aspect of truth. On the one hand, we have been studying for over a decade and a half the social hierarchies based on aggression and submission that are characteristic of many social groups. We know from personal observation, as well as from the literature, of nip orders in fishes, peck orders in flocks of several species of birds, and fighting orders in lizards and in mammals. Among these vertebrates there is usually one dominant animal that can bite, nip, or peck others without being attacked in return. Below it the others are ranked in various degrees of subservience. Similar dominance orders occur among such mammals as mice, cats, cows, monkeys, and men.

Those social hierarchies are based immediately on fighting or bluffing ability, on individual aggressiveness or meekness, as shown in pair contacts between the different members of the group. The order, once established, is not readily upset. With hens we have observed the same peck order to persist unchanged for as long as a year, and a year is a relatively long time in the life of a hen. These social orders are an expression of crude, person-against-person competition for social status. They furnish fair illustrations of the self-centered phase of group biology. Here is an aspect of the individualistic struggle for existence, and as such it illustrates an important phase of the Darwinian theory of evolution.

High position in the social peck order confers privileges. We know that top-ranking animals feed more freely; that high-ranking males of rhesus monkeys, sage grouse, and the common domestic fowl have more ready access to females. Low social status may lead to semistarvation, to psychological castration among cocks, to ejection from coveys of quail; among many species it forces the low-ranking birds into inferior territories. In some cases, high social rank carries responsibilities for leadership or for standing guard; in other instances no correlation with social services has as yet been demonstrated.

These studies of individual aggressiveness make one experimental approach to general and comparative sociology. Certain of the results obtained have been treated in some detail. This is the more spectacular, but fundamentally the less important, aspect of group biology.

Our second line of attack comes from a different quarter. For more than thirty years we have been experimenting with group-centered tendencies that long before my time were called cooperation. Careful students nowadays point out that, among lower organisms, proto-cooperation—as the beginnings of cooperation may properly be called—is entirely nonconscious. In this phase of our study we are investigating natural cooperation somewhat as many other biologists have been concerned with natural selection. Natural proto-cooperation in its simpler forms implies merely that the interrelations between cells, for example, are more beneficial than harmful for the individual, or that the interrelations between individuals are more beneficial than harmful for the given social group.

In the present book much time has been spent reviewing the types of modern evidence concerning the existence of such cooperation. Mere repetition does not necessarily make for acceptance, but a good purpose may be served by summarizing in outline form the types of modern evidence that I have found compelling.

1. At all levels of the animal kingdom, and under a variety of conditions, there is added safety in numbers up to a given point. Animals from the protozoans to insects or man meet many adverse conditions better if optimal numbers, rather than too few, are present. There

is danger also in overcrowding, but since that danger is better understood I put more emphasis on the danger of the population being too sparse—on the danger of undercrowding.

For certain animals mass protection exists when the organisms are exposed to heat. Mass protection from cold is more common, as is protection from many poisons and from other harmful chemicals. Optimal numbers also protect from ultra-violet radiation, from radical changes in salinity as from the sea to fresh water, and from many environmental deficiencies.

It is clear that all the evidence available could not be presented. There was not space to treat, for example, the evidence showing that separated cells of a sponge will not regenerate if too few are present and that the smallest embryonic grafts frequently fail to grow when somewhat larger ones succeed. I did, however, take space to show that if a natural population falls too low it is in danger of dying out even though theoretically able to persist.

2. In keeping with the relations just outlined, many organisms, both plants and animals, change or "condition" an unfavorable medium so that others following or associated with them can survive better and thrive when they could not do so in a raw, unconditioned medium.

3. Certain vital processes are favorably retarded by increased numbers up to a given density. For example, scattered spermatozoa of many aquatic organisms die more rapidly than they do when massed together. This is an interesting example of how competition may have cooperative end results. The massed spermatozoa are competing with each other for the limited amount of space available and as a result of this competition, they live longer. Competition is not always in opposition to cooperation; that

is, competition does not always produce disoperative effects.

4. Other biological processes are accelerated, perhaps beneficially, in the presence of many populations of optimal size and density. Such processes are slowed down both in oversparse populations and in those that are overcrowded. Cleavage rate in sea-urchin eggs follows this rule. Various kinds of protozoans show acceleration in rate of asexual reproduction in medium as contrasted with sparse population density, and similar phenomena may well have been a forerunner of

5. The evolution of the cooperative processes that are associated with sexual reproduction.

6. Colonial protozoa could hardly have evolved from solitary forms unless the simple colony of cells that remained attached after division had shown cooperative powers that were lacking when the individual cells were scattered singly. The evolution of many-celled animals from protozoans was probably based on similar relationships.

7. Each advance in complexity came from the natural selection of an increased ability in natural cooperation on the part of the evolving stock; the greater natural cooperation came first, then it was selected. Cooperation is a biological necessity both in individual evolution and in the evolution of groups.

8. Darwin in the *Origin of Species* recognized that a relatively large population is a highly important factor in natural selection. There is newer evidence that evolution proceeds most rapidly in populations of random-breeding organisms that are intermediate in size, as compared with similar populations that are overlarge or oversmall. Sewall Wright [140] has calculated that most rapid evolution is to be expected in a population that is much subdivided into

small interbreeding units interconnected by occasional emigrants.

9. The interdependence of organisms is shown by the repeated observation that all living forms, from the simplest to the most complex, live in ecological communities;[11] this is plainly seen in many closely knit groups such as oyster banks or coral reefs. Further, the evolution of truly social animals has occurred independently in widely separated divisions of the animal kingdom. These could hardly have arisen so frequently and from such diverse sources if a strong substratum of generalized natural cooperation, or at least of proto-cooperation, were not widespread among animals in nature. No free-living animal is solitary throughout its life history.

10. As with individual organisms, each advance in complexity of the social life of any group of animals is based on the development of some closer cooperation between the individual units of the evolving group.

We have good evidence, then, that these two types of social or subsocial interactions exist among animals: the self-centered, egoistic drives, which lead to personal advancement and self-preservation; and the group-centered, more or less altruistic drives, which lead to the preservation of the group, or of some members of it, perhaps at the sacrifice of many others.

The presence of egoistic forces in animal life has long been recognized. It is not so well known that the idea of the group-centered forces of natural cooperation also has a respectable history. Accordingly, I have presented a bare outline of the growth of this idea and mentioned some of its proponents, since many people are rightly interested in the men who have held an idea to be valid as well as in the supporting evidence.[6]

Widely dispersed knowledge concerning the important role of basic cooperative processes among living beings may lead to the acceptance of cooperation as a guiding principle both in social theory and as a basis for human behavior. Such a development when it occurs will alter the course of human history.

Literature Cited*

1. ALLEE, W. C. 1912. "An experimental analysis of the relation between physiological states and rheotaxis in Isopoda," *J. Exp. Zool.* 13:269-344.
2. ———. 1920. "Animal aggregations," *Anat. Rec.* 17: 340.
3. ———. 1931. *Animal Aggregations, a Study in General Sociology.* Chicago: University of Chicago press, 431 pp.
4. ———. 1934. "Recent studies in mass physiology," *Biol. Rev.* 9:1-48.
5. ———. 1937. "Evolution and behavior of the invertebrates," in *The World and Man as Science Sees Them*, edited by F. R. Moulton. Chicago: University of Chicago Press, pp. 294-346.
6. ———. 1943. "Where angels fear to tread: A con-

---

* No attempt has been made to document the text fully or to cite all of the important books and papers that have been consulted intensively. Many of these are cited in the bibliographies to be found in the literature here listed.

tribution from general sociology to human ethics," *Science* 97:514-25.

7. ———. 1945. "Biology and international relations," *New Republic* 112:816-17.

8. ———, AND BOWEN, EDITH. 1932. "Studies in animal aggregations: mass protection against colloidal silver among goldfishes," *J. Exp. Zool.* 61:185-207.

9. ———, BOWEN, EDITH, WELTY, J., AND OESTING, R. 1934. "The effect of homotypic conditioning of water on the growth of fishes, and chemical studies of the factors involved," *J. Exp. Zool.* 68:183-213.

10. ———, AND COLLIAS, N. 1940. "The influence of estradiol on the social organization of flocks of hens," *Endocrinology* 27:87-94.

11. ———, EMERSON, A. E., PARK, O., PARK, T., AND SCHMIDT, K. P. 1949. *Principles of Animal Ecology.* Philadelphia: Saunders, 837 pp.

12. ———, AND EVANS, GERTRUDE. 1937a. "Some effects of numbers present on the rate of cleavage and early development in *Arbacia*," *Biol. Bull.* 72:217-32. 1937b. "Further studies on the effect of numbers on the rate of cleavage in eggs of *Arbacia*," *J. Cell. and Comp. Physiol.* 10:15-28. 1937c. "Certain effects of numbers present on the early development of the purple sea-urchin, *Arbacia punctulata:* a study in experimental ecology," *Ecology* 18:337-45.

13. ———, AND FRANK, P. 1949. "The utilization of minute food particles by goldfish," *Physiol. Zool.* 22:346-58.

14. ———, FRANK, P., AND BERMAN, MARJORIE. 1946. "Homotypic and heterotypic conditioning in relation to survival and growth of certain fishes," *Physiol. Zool.* 19:243-58.

15. ———, AND MASURE, R. H. 1936. "A comparison of maze behavior in paired and isolated shell par-

rakeets (*Melopsittacus undulatus* Shaw) in a two-alley problem box," *J. Comp. Psych.* 22:131-56.

16. ———, OESTING, R., AND HOSKINS, W. 1936. "Is food the effective growth-promoting factor in homotypically conditioned water? *Physiol. Zool.* 9:409-32.

17. ———, AND ROSENTHAL, G. M., JR. 1949. "Group survival value for *Philodina roseola*, a rotifer," *Ecology* 30:395-97.

18. ———, AND WILDER, JANET. 1938. "Group protection for *Euplanaria dorotocephala* from ultra-violet radiation," *Physiol. Zool.* 12:110-35.

19. ALLPORT, F. H. 1924. *Social Psychology*. Boston: Houghton Mifflin, 453 pp.

20. ALVERDES, F. 1927. *Social Life in the Animal World*. New York: Harcourt, Brace, 216 pp.

21. ANDREWS, R. C. 1926. *On the Trail of Ancient Man:* a narrative of the field work of the Central Asiatic expeditions. New York: Putnam, 375 pp.

22. *The Auk*. 1932. "Notes and Comments." N. S. 49: 524.

23. BAILEY, V. 1931. *Mammals of New Mexico,* U. S. Dept. Agric. Bur. of Biol. Survey. N. Amer. Fauna, No. 53, 412 pp.

24. BAKER, O. E. 1937. "Human resources of the United States," *Science; Science News Suppl.* 86 (2223):12.

25. BALTZER, F. 1928. "Über metagame Geschlechtsbe-stimmung und ihre Beziehung zu einigen Problemen der Entwicklungsmechanik und Vererbung" (Zu-sammenfass. Schrift), *Verh. d. Deutsch. Zool. Ge-sellsch.* 32:273-325.

26. BATES, H. W. 1892. *The Naturalist on the River Amazons*. London: Murray, 395 pp.

27. BAYER, E. 1929. "Beiträge zur Zweikomponenten-theorie des Hungers," *Zeit. f. Psych.* 112:1-54.

28. BEEBE, W., HARTLEY, G., AND HOWES, P. 1916. *Tropical Wild Life in British Guiana.* New York: New York Zool. Soc., 504 pp.

29. BLATZ, W. C., MILLICHAMP, D., AND CHARLES, M. 1937. "The early social development of the Dionne quintuplets," *University of Toronto Studies. Child Development Series,* No. 13, 40 pp. Also in *Collected Studies on the Dionne Quintuplets.* University of Toronto Press.

30. BOHN, G., AND DRZEWINA, A. 1920. "Variations de la sensibilité à l'eau douce des *Convoluta* suivant les états physiologiques et le nombre des animaux en expérience," *C. R. Acad. Sci.* 171:1023-25.

31. CASTLE, G. B. 1934. "An experimental investigation of caste differentiation in *Zootermopsis augusticollis,*" in *Termites and Termite Control,* edited by C. A. Kofoid. Berkeley: University of California Press, 2nd edition, pp. 292-310.

32. CHAPMAN, F. M. 1935. "The courtship of Gould's manakin on Barro Colorado Island, Canal Zone," *Bull. Amer. Mus. Nat. Hist.* 68:471-523.

33. CHEN, S. C. 1938a. "Social modification of the activity of ants in nest building," *Physiol. Zool.* 10: 420-36. 1938b. "The leaders and followers among the ants in nest building," *ibid.* 10:437-55.

34. CHEVILLARD, L. 1935. Contribution à l'étude des échanges respiratoires de la Souris blanche adult. II. La température corporelle de la Souris et ses variations," *Ann. Physiol. et Physicochimie* 11:468-84.

35. CHILD, C. M. 1915. *Senescence and Rejuvenescence.* Chicago: University of Chicago Press, 481 pp.

36. ———. 1924. *Physiological Foundations of Behavior.* New York: Holt, 330 pp.

37. CHURCHMAN, J., AND KAHN, M. 1921. "Communal activity of bacteria," *J. Exp. Med.* 33:583-91.

38. CLEVELAND, L. R. 1934. "The wood-feeding roach, *Cryptocercus*, and its Protozoa, and the symbiosis between Protozoa and roach," *Mem. Amer. Acad. Arts and Sci.* 17:187-342.

39. COE, W. R. 1936. "Sexual phases in *Crepidula*," *J. Exp. Zool.* 72:455-77.

40. COMTE, A. 1830. *Cours de Philosophie Positive.* Paris: Schleicher, 6 vols.

41. CREW, F. A., AND MIRSKAIA, L. 1931. "The effects of density on an adult mouse population," *Biol. Gen.* 7:239-50.

42. DANSFORTH, C. H. 1934. "The interrelation of genetic and endocrine factors in sex," in *Sex and Internal Secretions,* edited by E. Allen. Baltimore: Williams and Wilkins, pp. 12-54.

43. DARLING, F. F. 1937. *A Herd of Red Deer.* London: Oxford University Press, 215 pp.

44. ———, 1938. *Bird Flocks and the Breeding Cycle.* London: Cambridge University Press, 124 pp.

45. DEEGENER, P. 1918. *Die Formen der Vergesellschaftung im Tierreiche. Ein systematisch-soziologischer Versuch.* Leipzig: Veit, 420 pp.

46. DOBZHANSKY, T. 1937. *Genetics and the Origin of Species.* New York: Columbia University Press, 364 pp.

47. DURKHEIM, E. 1902. *De la division du travail social.* Paris: Alcan, 460 pp.

48. ELLIS, H. 1929. *The Dance of Life.* Boston: Houghton Mifflin, 342 pp.

49. ESPINAS, A. V. 1878. *Des sociétés animales.* Paris: Baillière, 582 pp.

50. FALK, ISADORE. 1927. "Does infant welfare operate

220

to preserve the unfit?" *Amer. J. Pub. Health* 17:142-47.

51. FEUERBACH, L. 1846-1890. *Sämmtliche Werke.* Leipzig: Wigand, 10 vols.

52. FISCHEL, W. 1927. "Beiträge zur Soziologie des Haushuhns," *Biol. Zentralbl.* 47:678-96.

53. FOLKS, H. 1920. *The Human Costs of the War.* New York: Harper, 326 pp.

54. FORBES, S. A. 1887. "The lake as a microcosm," *Bull. Peoria Acad. Sci.* Reprinted in *Ill. State Nat. Hist. Survey Bull.* 15:537-50.

55. FORBUSH, E. H. 1925-1929. *Birds of Massachusetts and Other New England States.* Mass. Dept. Agric., Vols. 1-3.

56. FOWLER, J. R. 1931. "The relation of numbers of animals to survival in toxic concentrates of electrolytes," *Physiol. Zool.* 4:214-45.

57. GARNER, M. R. 1934. "The relation of numbers of *Paramecium caudatum* to their ability to withstand high temperatures," *Physiol. Zool.* 7:408-34.

58. GATES, MARY, AND ALLEE, W. C. 1933. "Conditioned behavior of isolated and grouped cockroaches on a simple maze," *J. Comp. Psych.* 15:331-58.

59. GAUSE, G. F. 1934. *The Struggle for Existence.* Baltimore: Williams and Wilkins, 163 pp.

60. GEDDES, P., AND THOMPSON, J. A. 1911. *Evolution.* New York: Holt, 256 pp.

61. GOULD, H. N. 1917a. Studies on sex in the hermaphrodite mollusc *Crepidula plana.* I. "History of the sexual cycle," *J. Exp. Zool.* 23:1-69. 1917b. II. "Influence of environment on sex," *ibid.* 23:225-50. 1919. III. "Transference of the male-producing stimulus through sea-water," *ibid.* 29:113-20.

62. GRAVE, B. H., AND DOWNING, R. C. 1928. "The

longevity and swimming ability of spermatozoa,"
J. Exp. Zool. 51:383-88.

63. GROSS, A. O. 1928. "The Heath Hen," *Mem. Bost. Soc. Nat. Hist.* 6:491-588.

64. GUHL, A. M., COLLIAS, N. E., AND ALLEE, W. C. 1945. "Mating behavior and the social hierarchy in small flocks of white Leghorns," *Physiol. Zool.* 18: 365-90.

65. GULICK, A. 1905. "Evolution, racial and habitudinal," *Pub. Carnegie Inst.* 25:1-269.

66. HANKINS, F. H. 1937. "German policies for increasing births," *Amer. J. Soc.* 42:630-52.

67. HARLOW, H. F. 1932. "Social facilitation of feeding in the albino rat," *J. Genet. Psych.* 41:211-21.

68. HARNLY, M. H. 1929. "An experimental study of environmental factors in selection and population," *J. Exp. Zool.* 53:141-70.

69. HICKS, F. 1920. *The New World Order.* New York: Doubleday Page, 496 pp.

70. HOGG, J. 1854. "Observations on the development and growth of the water snail (*Lymnaeus stagnalis*)," *Quart. J. Micros. Sci.* 2, in *Trans. Micros. Soc.* 2:91-103.

71. HOLMES, S. J. 1921. *The Trend of the Race:* a study of present tendencies in the biological development of civilized mankind. New York: Harcourt, Brace, 396 pp.

72. ———. 1936. *Human Genetics and Its Social Import.* New York: McGraw-Hill, 414 pp.

73. HOWARD, H. E. 1920. *Territory in Bird Life.* London: Murray, 308 pp.

74. HUDELSON, E. 1928. *Class Size at the College Level.* Minneapolis: University of Minnesota Press, 299 pp. 1932. "Class size at the college level," *No. Cent. Assoc. Quart.* 6:371-81.

75. HUNT, H. R. 1930. *Some Biological Aspects of War.* New York: Galton, 118 pp.

76. JOHNSON, W. H. 1937. "Experimental populations of microscopic organisms," *Amer. Nat.* 71:5-20.

77. JONES, F. M. 1931. "The gregarious sleeping habits of *Heliconius charithonia* L." *Proc. Ent. Soc. London* 6:4-10.

78. JORDAN, D. S. 1903. *The Blood of the Nation:* a study of the decay of races through the survival of the unfit. Boston: Amer. Unitar. Assoc., 82 pp. 1907. *The Human Harvest:* a study of the decay of nations through the survival of the unfit. Boston: Amer. Unitar. Assoc., 122 pp. 1915. *War and the Breed:* the relation of war to the downfall of nations. Boston: Beacon Press. 265 pp.

79. ———, AND JORDAN, H. E. 1914. *War's Aftermath.* Boston: Houghton Mifflin, 103 pp.

80. KATZ, D., AND TOLL, A. 1923. "Die Messung von Charakter- und Begabungs-unterschieden bei Tieren (Versuch mit Hühnern)," *Zeit. f. Psych.* 93:287-311.

81. KEITH, A. 1949. *A New Theory of Evolution.* New York: Philosophical Library, 451 pp.

82. KELLOGG, V. L. 1912. *Beyond War:* a chapter in the natural history of man. New York: Holt, 172 pp.

83. KROPOTKIN, P. 1914. *Mutual Aid, a Factor in Evolution.* New York: Knopf, 2nd edition, 223 pp.

84. LANGE, F. A. 1925. *The History of Materialism.* New York: Harcourt, Brace, 3 vols. (Original German edition, 1865.)

85. LIGHT, S. F., 1942-43. The determination of the castes of social insects. *Quart. Rev. Biol.* 17:312-326; 18:46-63.

86. LIVENGOOD, W. 1937. "An experimental analysis of certain factors affecting growth of goldfishes in

homotypically conditioned water," *Copeia* 2:81-88.

87. MACLAGEN, D. S. 1932. "The effect of population density upon rate of reproduction with special reference to insects," *Proc. Roy. Soc. B* 111:437-54.

88. MALINOWSKI, B. 1937. "Culture as a determinant of behavior," in Harvard tercentenary conference, *Factors Determining Human Behavior*. Cambridge: Harvard University Press, 168 pp.

89. MAST, S. O., AND PACE, D. 1946. The nature of the growth-promoting substance produced by *Chilomonas paramecium*. *Physiol. Zool.* 19:223-235.

90. MASURE, R. H., AND ALLEE, W. C. 1934. "The social order in flocks of the common chicken and the pigeon," *Auk* 51:306-27.

91. MATTHEWS, L. H. 1932. "Lobster-krills; anomuran Crustacea that are the food of whales," *Discovery Reports, Gov. Dependencies, Falkland Islands* 5:467-84.

92. MÖBIUS, K. 1883. "The oyster and oyster culture," *U. S. Comm. Fish and Fisheries. Report. 1880. Part VIII:* 683-751.

93. MONTAGU, A. 1950. *On Being Human*. New York: Schuman, 125 pp.

94. MURCHISON, C. 1935a. The experimental measurement of a social hierarchy in *Gallus domesticus*. I. "The direct identification and direct measurement of social reflex No. 1 and social reflex No. 2," *J. Gen. Psych.* 12:3-39. 1935b. II. "The identification and inferential measurement of social reflex No. 1 and social reflex No. 2 by means of social discrimination," *J. Soc. Psych.* 6:3-30. 1935c. III. "The direct and inferential measurement of social reflex No. 3," *J. Genet. Psych.* 46:76-102. 1935d. IV. "Loss of body weight under conditions of mild starvation as a function of social dominance," *J. Gen.*

Psych. 12:296-312. 1935e. V. "The post-mortem measurement of anatomical features," *J. Soc. Psych.* 6:172-81.

95. MURPHY, G., AND MURPHY, LOIS. 1931. *Experimental Social Psychology.* New York: Harper, 709 pp.

96. NICHOLS, J. T. 1931. "Notes on the flocking of shore birds," *Auk* 48:181-85.

97. OESTING, R., AND ALLEE, W. C. 1935. "Further analysis of the protective value of biologically conditioned fresh water for the marine turbellarian, *Procerodes wheatlandi.* IV. The effect of calcium," *Biol. Bull.* 68:314-26.

98. ORTON, J. H. 1909. "On the occurrence of protandric hermaphroditism in the mollusc *Crepidula fornicata,*" *Proc. Roy. Soc. B* 81:468-84.

99. PARK, T. 1932. "Studies in population physiology: the relation of numbers to initial population growth in the flour beetle, *Tribolium confusum* Duval," *Ecology* 13:172-82.

100. ———. 1933. "Studies in population physiology: II. Factors regulating initial growth of *Tribolium confusum* populations," *J. Exp. Zool.* 65:17-42.

101. PATTEN, W. 1920. *The Grand Strategy of Evolution.* Boston: Badger, 429 pp.

102. PEEBLES, FLORENCE. 1929. "Growth-regulating substances in Echinoderm larvae," *Biol. Bull.* 57:176-87.

103. PEARL, R. 1920. "The effect of war on the chief factors of population change," *Science* 51:553-56.

104. ———. 1921. "A further note on war and population," *Science* 53:120-21.

105. ———. 1936. "War and overpopulation," *Curr. Hist.* 43:589-94.

106. ———, AND GOULD, SOPHIA. 1936. "World population growth," *Human Biol.* 8:399-419.

107. ———, AND PARKER, S. 1922. "On the influence of

density of population upon the rate of reproduction in *Drosophila,*" *Proc. Nat. Acad. Sci.* 8:212-19.

108. PHILLIPS, J. Personal communication.

109. POPENOE, P. B., AND JOHNSON, R. 1918. *Applied Eugenics.* New York: Macmillan, 429 pp.

110. RABAUD, E. 1937. *Phénomène social et sociétés animales.* Paris: Alcan, 321 pp.

111. REICH, K. 1938. Studies on the physiology of Amoeba. *Physiol. Zool.* 11:347-358.

112. REIGHARD, J. 1893. "The ripe eggs and spermatozoa of the wall-eyed pike," *Bien. Rept. Mich. State Board Fish Comm.* 10:93-171.

113. ———. 1920. "The breeding behavior of the suckers and minnows. I. The suckers," *Biol. Bull.* 38:1-32.

114. REINHARD, H. J. 1927. "The influence of parentage, nutrition, temperature and crowding on wing production in *Aphis gossypii,*" *Texas Agric. Exp. Sta. Bull.* 353:5-19.

115. RETZLAFF, E. 1939. Studies in mass physiology; growth rate with the white mouse. *J. Exp. Zool.* 81:343-56.

116. ———. 1938. "Studies in population physiology with the albino mouse," *Biol. Gen.* 14:238-65.

117. ROBERTSON, T. B. 1921. "Experimental studies on cellular multiplication. II. The influence of mutual contiguity upon reproductive rate and the part played therein by the 'X-substance' in bacterised infusions which stimulate the multiplication of Infusoria," *Biochem. J.* 15:612-19.

118. SCHJELDERUP-EBBE, T. 1922. "Beiträge zur Sozialpsychologie des Haushuhns," *Zeit. f. Psych.* 88:225-52.

119. ———. 1931. "Die Despotie im sozialen Leben der Vögel," Thurnwald, *Forschungen zur Völkerpsychologie und Sozialogie* 10 (2):77-140.

226

120. ———. 1935. "Social behavior of birds," in *A Hand-book of Social Psychology*, edited by C. Murchison. Worcester: Clark University Press, pp. 947-72.

121. SCHNEIRLA, T. C. 1946. "Problems in the biopsychology of social organization," *J. Abnormal and Soc. Psych.* 41:385-402.

122. SELOUS, E. 1931. *Thought-transference (or What?) in Birds.* New York: Smith, 255 pp.

123. SHAW, GRETCHEN. 1932. "The effect of biologically-conditioned water upon rate of growth in fishes and Amphibia," *Ecology* 13:263-78.

124. SHOEMAKER, H. 1939. Social hierarchy in flocks of the canary. *The Auk.* 56:381-406.

125. Society for the Psychological Study of Social Issues. 1937. *Bulletin* 2:11-12.

126. *The Statesman's Year Book.* 1922-1936. London: Macmillan.

127. SZYMANSKI, J. S. 1912. "Modification of the innate behavior of cockroaches," *J. An. Behav.* 2:81-90.

128. UVAROV, B. P. 1928. *Locusts and Grasshoppers.* London: Imp. Bur. Entom., 352 pp.

129. VETULANI, T. 1931. "Untersuchungen über das Wachstum der Säugetiere in Abhängigkeit von der Anzahl zusammengehaltener Tiere. I. Teil Beobachtungen an Mäusen," *Biol. Gen.* 7:71-98.

130. WELTY, J. C. 1934. "Experiments in group behavior of fishes," *Physiol. Zool.* 7:85-128.

131. WHEELER, W. M. 1913. *Ants, Their Structure, Development and Behavior.* New York: Columbia University Press, 663 pp.

132. ———. 1923. *Social Life Among the Insects:* being a series of lectures delivered at the Lowell Institute in Boston. New York: Harcourt, Brace, 375 pp.

133. ———. 1928. *The Social Insects, Their Origin and Evolution.* New York: Harcourt, Brace, 378 pp.

134. ——. 1930. "Societal evolution," in *Human Biology and Racial Welfare*, edited by E. Cowdry. New York: Hoeber, pp. 139-55.

135. ——. 1933. Address before American Society of Naturalists. Cambridge, 1933.

136. ——. 1937. *Mosaics and Other Anomalies Among Ants*. Cambridge: Harvard University Press, 95 pp.

137. WHELPTON, P. K. 1935. "Why the large rise in the German birth rate?" *Amer. J. Soc.* 41:299-313.

138. WOLFSON, C. 1935. "Observations on *Paramecium* during exposure to sub-zero temperatures," *Ecology*, 16:630-39.

139. WRIGHT, S. 1931. "Evolution in Mendelian populations," *Genetics* 16:97-159. 1932. "Roles of mutation, inbreeding, crossbreeding and selection in evolution," *Proc. 6th. Int. Congress Genetics* 1:356-66.

140. ——. 1945. "Tempo and mode in evolution. A critical review," *Ecology* 26:415-19.

141. WRIGHT, Q. 1942. *A Study of War*." Chicago: University of Chicago Press, 2 vols., 1552 pp.

142. ZELLER, E. 1931. *Outlines of the History of Greek Philosophy*. New York: Harcourt, Brace, 324 pp.

Index

Aggregations, in nature, 17; hibernating, 17; breeding, 17; migrating, 18, 19; colonial animals, 23; forced, 24; feeding, 25; overnight, 27, 157; relation to social life, 28, 174, 209

Alcohol, mass protection from, 32

Alverdes, F., 13

Ancestral tree of animals, 58

Andrews, Roy C., 20

Antelope, 20

Ants, 17; effect of numbers on digging, 101; importance of females, 165; castes, 167

Aphids, 174

Appetite, social, 26, 28, 156

Arbacia eggs, 45; spermatozoa, 47, 56; effect of numbers on rate of cleavage, 47; effect of extracts, 49

Bacteria, mass protection, 43; food for protozoa, 52

Baker, O. E., 182

Bats, 20

Bavaria, population trend, 185, 187

Beebe, William, 157

Bees, 165, 167; solitary, 27; importance of females, 165; castes, 167

Beetles, hibernation, 17

Belgium, population, 189

Birds, 28, 29, 60, 77, 96, 112, 129, 157, 166

Birthrate, 183, 186

Bison, 21

Bonellia, 163

Butterflies, overnight aggregations, 27

Calcium, protective value, 43

Canaries, social order, 141, 143, 152

Caribou, 20

Caste, 159, 166, 175

Castle, G. B., 170

Chapman, Frank M., 97